Math Kangaroo in USA
Levels 3 and 4

Questions and Solutions

1998–2019

© Copyright 2020 by Math Kangaroo in USA, NFP, Inc.
www.mathkangaroo.org

Printed by:
Classic Printing & Thermography
Wood Dale, IL

This book contains the math problems (questions) that were solved by Math Kangaroo participants in the years 1998-2019, with solutions. It can be used at home and at school.

For additional copies of this book, please contact the publisher:
Math Kangaroo in USA, NFP
www.mathkangaroo.org
info@mathkangaroo.org

ISBN 978-0-9814868-3-3

Preface

After the book for students of grades one and two was published, we received questions about books for older children.

Here is the answer to the requests from parents: *Math Kangaroo in USA Levels 3 and 4, Questions and Solutions, 1998–2019*.

The book contains mathematical problems solved by Math Kangaroo participants of grades 3 and 4 since the competition's first year in the USA through 2019.

Math Kangaroo is an international competition in mathematics prepared and managed by *Association Kangourou sans Frontières* (www.aksf.org) since 1991. The USA joined the organization several years later.

Questions are prepared and proposed by the *Association* at its annual meetings where representatives from around eighty member countries gather to select mathematical problems which are unique and appealing to young participants. There are twelve levels of participation: from grade 1 to grade 12 (senior class of high school).

This book was prepared by our dedicated contributors: Dave Zarach, who collected all the question sets and verified the solutions, Andrzej Zarach, Ph.D., who oversaw his son's project, and Ela Zarach, who facilitated the work. Joanna Matthiesen worked closely with them and contributed to every step of the process.

Agata Gazal took care of editorial work. Several members of the Math Kangaroo in USA Development Team contributed to the development of solutions.

We hope the book is well received by students and teachers and it contributes to their efforts of mastering mathematical skills.

Enjoy the book!

Maria Omelanczuk
Math Kangaroo in USA
September 2020

Association Kangourou sans Frontières
after selecting questions for
Math Kangaroo 2020,
Chicago, October 16–20, 2019

TABLE OF CONTENTS

Part I: Questions 7

Problems from Year 1998 9
Problems from Year 1999 12
Problems from Year 2000 15
Problems from Year 2001 19
Problems from Year 2002 23
Problems from Year 2003 27
Problems from Year 2004 31
Problems from Year 2005 35
Problems from Year 2006 39
Problems from Year 2007 43
Problems from Year 2008 47
Problems from Year 2009 51
Problems from Year 2010 55
Problems from Year 2011 59
Problems from Year 2012 63
Problems from Year 2013 67
Problems from Year 2014 71
Problems from Year 2015 75
Problems from Year 2016 79
Problems from Year 2017 83
Problems from Year 2018 88
Problems from Year 2019 93

Part II: Solutions 99

Solutions for Year 1998 101
Solutions for Year 1999 104
Solutions for Year 2000 107
Solutions for Year 2001 112
Solutions for Year 2002 115
Solutions for Year 2003 119
Solutions for Year 2004 123
Solutions for Year 2005 128
Solutions for Year 2006 131
Solutions for Year 2007 135
Solutions for Year 2008 141
Solutions for Year 2009 147
Solutions for Year 2010 153
Solutions for Year 2011 156
Solutions for Year 2012 161
Solutions for Year 2013 166
Solutions for Year 2014 169
Solutions for Year 2015 175

Solutions for Year 2016 181
Solutions for Year 2017 185
Solutions for Year 2018 191
Solutions for Year 2019 196

Part III: Answers 201

Answer Keys 203

Part I: Questions

Problems from Year 1998 (21 problems)

Problems 3 points each

1. At the zoo, Bob saw kangaroos for the first time. He noticed that each kangaroo had four legs, two ears, and one tail. For fun, he counted all the legs, ears and tails, and got the number 63. How many kangaroos did he see?

   (A) 6 (B) 7 (C) 9 (D) 10 (E) 12

2. John wanted to place parentheses in the expression $6 \times 8 + 20 \div 4 - 2$ in such a way as to get 58 as the result. Which of the ways shown below would give him this result?

   (A) $6 \times (8 + 20) \div 4 - 2$ (B) $(6 \times 8 + 20 \div 4) - 2$ (C) $(6 \times 8 + 20) \div 4 - 2$
   (D) $6 \times 8 + 20 \div (4 - 2)$ (E) $6 \times (8 + 20 \div 4) - 2$

3. How many triangles can you see in the picture?

   (A) 2 (B) 6 (C) 8 (D) 10 (E) 12

   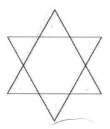

4. Mary lives in a tall building in apartment number 17. There are stores on the first floor. Above the stores, there are 3 apartments on each floor, numbered consecutively. On what floor does Mary live?

   (A) 4th (B) 5th (C) 6th (D) 7th (E) 9th

5. What number do we need to place inside □ to make $12 \times 12 \times 12 = 6 \times □ \times 6$ true?

   (A) 12 (B) 24 (C) 48 (D) 72 (E) 60

6. How many three-digit numbers can you create using the digits 3, 0, and 7, and using each digit only once?

   (A) 2 (B) 3 (C) 4 (D) 5 (E) 6

7. Out of how many blocks is this tower built?

   (A) 20 (B) 22 (C) 25 (D) 28 (E) 30

   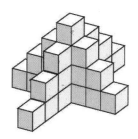

Problems 4 points each

8. A kangaroo is traveling from START to FINISH using the paths shown in the picture. Each segment is marked with the time (in minutes) which the kangaroo needs to travel that segment. What is the shortest time needed for the kangaroo to reach FINISH?

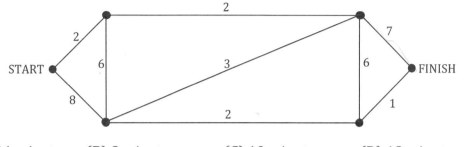

(A) 11 minutes   (B) 8 minutes   (C) 10 minutes   (D) 18 minutes   (E) 6 minutes

9. Joanna baked some cookies. She tried to divide them evenly, first between 2 plates, then between 3 plates, and finally between 4 plates. Each time she had one cookie left over. What is the smallest number of cookies Joanna could have baked?

(A) 9   (B) 10   (C) 11   (D) 12   (E) 13

10. I chose a certain number. I then subtracted 40 from it. Then I added 2000 and as a result I now have 3250. What number did I choose in the beginning?

(A) 2040   (B) 1960   (C) 1290   (D) 3210   (E) 1250

11. On Monday morning, a snail fell down a well which is 5 meters deep. During the day, it climbs up 2 meters, but during the night it slides down 1 meter. On what day of the week will the snail get out of the well?

(A) Tuesday   (B) Wednesday   (C) Thursday   (D) Friday   (E) Monday

12. Adam cut five identical square sheets of paper into two pieces. From which of the five pieces below was the piece marked with Z cut?

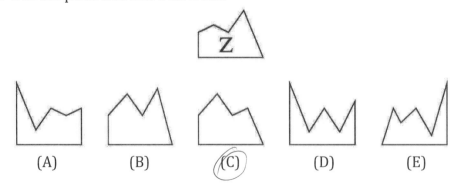

13. There were 31 runners competing in a race. The number of runners who finished before John is four times smaller than the number of runners who finished later than John. In what place did John finish?

    (A) 6   (B) 7   (C) 8   (D) 20   (E) 21

14. Half a loaf of bread costs 6 pence more than one-fourth of a loaf of bread. How many pence does a whole loaf of bread cost? (Note: A pence is an English coin.)

    (A) 6   (B) 12   (C) 18   (D) 24   (E) 30

Problems 5 points each

15. Write the numbers into the blank boxes in the pyramid to the right according to the pattern shown below.

       $4 = \dfrac{3+5}{2}$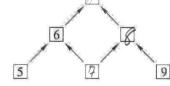

    What number is at the top of the pyramid?

    (A) 5   (B) 7   (C) 8   (D) 9   (E) 12

16. There are 15 balls in a box: white balls, red balls, and black balls. The number of white balls is 7 times greater than the number of red balls. How many black balls are there in the box?

    (A) 1   (B) 3   (C) 5   (D) 7   (E) 9

17. Paul was going to buy 4 servings of ice cream, but he was 80 cents short. So, he bought 3 servings and had 30 cents left. What was the price of one serving of ice cream?

    (A) 70 cents   (B) 80 cents   (C) 90 cents   (D) 1 dollar   (E) 1 dollar and 10 cents

18. We are making a square "chessboard" using matches that are 5 centimeters long. One side of the chessboard will be 1 meter long. The picture shows the upper left-hand corner of the board. How many matches will we use? (Note: 1 m = 100 cm.)

    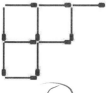

    (A) 400   (B) 480   (C) 640   (D) 840   (E) 960

19. How many three-digit numbers are there that have the sum of their digits equal to 5? (For example, 122 is such a number, because 1 + 2 + 2 = 5.)

    (A) 10   (B) 15   (C) 20   (D) 25   (E) 30

20. Ann is 3 years older than Barb and 2 years younger than Cali. Dorothy is 1 year younger than Barb. How much older is Cali than Dorothy?

    (A) 5 years   (B) 6 years   (C) 4 years   (D) 2 years   (E) They are the same age.

21. In a certain soccer tournament, the winning team gets 3 points, the losing team gets 0 points, and in the case of a tie both teams get 1 point each. My team played 31 games and received 64 points. 7 of the games were ties. How many games did my team lose?

    (A) 0   (B) 5   (C) 19   (D) 21   (E) 24

Problems from Year 1999

Problems 3 points each

1. Beata has two dolls, three apples, one chocolate bar, two oranges, five peaches and one bike. How many pieces of fruit does Beata have?

    (A) 3   (B) 5   (C) 10   (D) 18   (E) 21

2. What number is in the part that is common to four circles (see the picture)?

    (A) 5   (B) 9   (C) 7   (D) 4   (E) 6

3. In how many places do we need to break a wooden stick in order to get 5 pieces?

    (A) 3   (B) 4   (C) 5   (D) 60   (E) It depends on how long the stick is.

4. Karl is now 10 years old, and Alice is 3 years old. How many years from now will Karl be twice as old as Alice?

    (A) 5   (B) 10   (C) 4   (D) 1   (E) 3

5. Anna and her sister Barbara go to the same school, but take two different ways to get to school. Whose way is shorter?

    (A) Anna's way   (B) Barbara's way   (C) It depends on the distance to the school.
    (D) Both ways have the same length.   (E) It is impossible to determine.

6. Our class has 30 students. The number of boys is four times greater than the number of girls. How many girls are there in our class?

   (A) 24 (B) 16 (C) 12 (D) 8 (E) 6

7. How much does the orange weigh (see the picture)?

   (A) 200 g (B) 205 g (C) 155 g (D) 5 g
   (E) It cannot be determined.

8. The cat says, "The length of my tail is 12 cm and half the length of my tail." How long is the cat's tail?

   (A) 18 cm (B) 24 cm (C) 12 cm (D) 9 cm (E) 6 cm

Problems 4 points each

9. My mom's birthday is on a Sunday, and my dad's birthday is 55 days later. On what day of the week will my dad's birthday be?

   (A) Sunday (B) Monday (C) Tuesday (D) Thursday (E) Saturday

10. Two basketball teams are playing against each other. The first team to win four games wins the tournament. There are no ties. What is the greatest number of games that could take place after which the winner will be known for sure?

    (A) 8 (B) 7 (C) 6 (D) 5 (E) 4

11. Instead of adding 27 to a certain number, John subtracted 27 from that number. What is the difference between John's result and the result he should have gotten?

    (A) 27 (B) 0 (C) 54 (D) 100 (E) 3

12. A goldsmith decided to cut a golden cube with an edge of 4 cm into small cubes each with an edge of 1 cm. How many small cubes will he have?

    (A) 64 (B) 48 (C) 32 (D) 16 (E) 12

13. A pail filled with milk to the top weighs 25 kilograms, and a pail filled half-way weighs 13 kilograms. How much does an empty pail weigh?

    (A) 2 kilograms (B) ½ kilogram (C) 1 ½ kilograms
    (D) 1 kilogram (E) 2 ½ kilograms

14. In Grandma's pantry, there is a jar with 650 g of jam. Each day her grandson Tom eats 5 teaspoons of jam from the jar. Each teaspoon holds 6 g of jam. How much jam will there be left in the jar after 20 days?

(A) 50 g   (B) 530 g   (C) 550 g   (D) 1250 g   (E) The jar will be empty.

15. Each of the kangaroo's eleven children has eleven children, and each of them also has eleven children. How many great-grandchildren does the kangaroo have?

(A) 111   (B) 121   (C) 11211   (D) 1331   (E) 12321

16. What is the least possible number of children in the Kowalski family if each of the children has at least one brother and at least one sister?

(A) 1   (B) 2   (C) 3   (D) 4   (E) 5

Problems 5 points each

17. Peter opened a book and found that the sum of the page number on the left and the page number on the right is equal to 21. What is the product of the two page numbers?

(A) 121   (B) 100   (C) 420   (D) 110   (E) 426

18. Father Virgil is taking care of 143 children. Each day, each child gets half a liter of milk with breakfast. The milk from one cow is enough for 40 children. What is the least number of cows that Father Virgil needs to have?

(A) 2   (B) 3   (C) 4   (D) 5   (E) 6

19. A kangaroo wants to make a rectangular bedspread 1.5 m long and 1 m wide using square scraps which measure 10 cm × 10 cm. At every point where four squares meet she wants to place a fancy button. How many buttons will she need?

(A) 150   (B) 104   (C) 126   (D) 140   (E) 135

20. Pinocchio's wooden nose is 3 cm long. Whenever Pinocchio lies, the length of his nose doubles. How long will his nose be after he tells 6 lies?

(A) 192 cm   (B) 67 cm   (C) 96 cm   (D) 18 cm   (E) 384 cm

21. In the yard there is an equal number of pigs, ducks, and chickens. Together, they have 144 legs. How many ducks are there in the yard?

(A) 18   (B) 21   (C) 35   (D) 42   (E) 43

PROBLEMS 1999

22. One number was chosen from the numbers 51, 52, 53, 54 and 55, and the digit 0 was placed between the digits of that number. What is the difference between the new number and the number which was chosen?

(A) 500 (B) 50 (C) 550 (D) 450
(E) The difference depends on which number was chosen.

23. If Grandma gave each of her grandchildren 10 pieces of candy, there would not be enough candy for one of the grandchildren. If she gave each one of them 8 pieces of candy, she would have 6 pieces of candy left. How many grandchildren does she have?

(A) 4 (B) 6 (C) 8 (D) 10 (E) 12

24. The figure shown rotates clockwise and makes one full rotation in one hour. Its position at 12:00 p.m. it is shown in the picture to the right. What will it look like at 2:15 p.m.?

(A) (B) (C) (D) (E)

Problems from Year 2000

Problems 3 points each

1. A birthday candle stays lit for 15 minutes. For how long will 10 birthday candles stay lit if they are lit at the same time and no one blows them out?

(A) 1.5 minutes (B) 15 minutes (C) 150 minutes (D) 1.5 hours (E) 15 hours

2. The little kangaroo was sick. Dr. Ohpain prescribed 3 pills for him to take one at a time every twenty minutes. How many minutes after taking the first pill will the little kangaroo take the last pill?

(A) 20 (B) 30 (C) 40 (D) 50 (E) 60

3. For which of the following numbers is the product of the digits greater than the sum of the digits?

(A) 112 (B) 209 (C) 312 (D) 222 (E) 211

PROBLEMS 2000

4. Gavel lives on the second floor, and Pavel lives in the same building but has to walk up twice as many stairs as Gavel. There are no stairs to the entrance of the building. On which floor does Pavel live?

   (A) on the 2nd floor (B) on the 3rd floor (C) on the 4th floor
   (D) on the 5th floor (E) on the 6th floor

5. Four candy bars and three lollipops cost $4.50. One candy bar costs 90 cents. How much does one lollipop cost?

   (A) 20 cents (B) 30 cents (C) 40 cents (D) 50 cents (E) 60 cents

6. One tour bus can seat no more than 55 people. What is the smallest number of buses needed to seat 160 people?

   (A) 1 (B) 2 (C) 3 (D) 4 (E) 5

7. A person needs 12 minutes to walk around a square plaza. How much time will it take for the same person to walk at the same pace around a plaza that has an area that is four times greater?

   (A) 48 minutes (B) 24 minutes (C) 30 minutes (D) 20 minutes (E) 36 minutes

8. It is 120 km from Zakopane to the Krakow airport. Buses leave from Zakopane for the airport at 30 minutes past every hour. The buses drive with an average speed of 60 km per hour. A group of "Kangaroos," members of a math camp in Zakopane, is supposed to arrive at the airport at 11:30. What is the latest time their bus had to leave Zakopane to get them to the airport on time?

   (A) 7:30 (B) 8:30 (C) 9:30 (D) 10:30 (E) 11:30

Problems 4 points each

9. During the time that Kathy eats two bowls of ice cream, Betty eats three bowls of ice cream. The two girls ate 10 bowls of ice cream in one hour. How many bowls of ice cream did Kathy eat?

   (A) 3 (B) 4 (C) 5 (D) 6 (E) 7

10. Which four digits need to be removed from the number 4921508 to get the smallest possible three-digit number?

    (A) 4, 9, 2, 1 (B) 4, 2, 1, 0 (C) 1, 5, 0, 8 (D) 4, 9, 2, 5 (E) 4, 9, 5, 8

© Math Kangaroo in USA, NFP www.mathkangaroo.org

11. In each of two baskets there were 12 apples. Aria took a certain number of apples from the first basket. From the second basket, Zoe took a number of apples equal to the number of apples remaining in the first basket. How many apples were left in both baskets altogether?

(A) 6   (B) 12   (C) 18   (D) 20   (E) 24

12. Students walked to the museum in rows of three. Adam, Bart, and Carl noticed that they were in the seventh row from the front and in the fifth row from the back. How many students went to the museum?

(A) 12   (B) 24   (C) 30   (D) 33   (E) 36

13. Fourteen cats took part in a cat play. Some of them played moms and some of them played their kids. Every mom in this play had at least two kids. What is the greatest possible number of cat-moms in the play?

(A) 3   (B) 4   (C) 5   (D) 6   (E) 7

14. The first and the second scales are balanced (see the picture). How many plums do you need to place on the left side of the third scale to keep it in balance?

(A) 2   (B) 3   (C) 4   (D) 5   (E) 6

15. Each of the five neighbors owns a rectangular plot of land with the same area. The parts of the land with flowers growing on them are fenced in (solid line in the pictures). Who has the longest fence?

(A) Mr. Adam   (B) Mr. John   (C) Mr. Jack   (D) Mr. Peter   (E) Mr. Mark

16. For his birthday Patrick got a box with some identical cube blocks. He used all of them to make two projects (see the picture). All the blocks together weigh 900 grams. The project on the left weighs 300 grams and the picture shows all the blocks it is made of. How many blocks in the figure on the right are not shown?

(A) 4   (B) 5   (C) 6   (D) 7   (E) 8

Problems 5 points each

17. Altogether, 6 hens eat 8 cups of grain in 3 days. How many cups of grain will 3 hens eat in 9 days?

(A) 10 (B) 12 (C) 14 (D) 16 (E) 9

18. Adrianna's birthday present is placed in a box with dimensions of 10 cm × 10 cm × 30 cm and wrapped with a ribbon as shown in the picture. What is the length of the ribbon?

(A) 200 cm (B) 240 cm (C) 260 cm (D) 300 cm (E) 250 cm

19. Three kangaroos were born consecutively every 4 years. Right now, the oldest kangaroo is 5 times as old as the youngest one. How old is the youngest kangaroo?

(A) 10 (B) 8 (C) 6 (D) 4 (E) 2

20. When Mary was leaving home between 8 and 9 o'clock in the morning, she noticed that the hour hand and the minute hand on her watch were overlapping. When she returned home between 2 and 3 o'clock in the afternoon, the hour hand and the minute hand formed a straight line (see the picture). How long was Mary away from home?

(A) 5 hours (B) 5 and a half hours (C) 6 hours (D) 6 and a half hours (E) 7 hours

21. Find the number which has the following property: If we add this number and half of this number, we will get a number which is 3 less than twice the original number.

(A) 2 (B) 4 (C) 6 (D) 8 (E) 10

22. Three identical dice are placed one on top of the other (see the picture). The sides which touch each other have the same numbers of dots. What is the number of dots on the bottom of the lowest die?

(A) 1 (B) 2 (C) 3 (D) 5 (E) 6

23. Pete wanted to draw the picture of a kangaroo shown, without lifting his pencil from the paper and without going over the same line twice. At what point should he start (see the picture)?

(A) A (B) B or C (C) D or E (D) K
(E) There is no such point; this is impossible.

24. A magical ball falling to the ground bounces twice as high as the height from which it was dropped. From what height was the ball dropped if it reached the height of 320 cm after the second bounce?

(A) 80 cm   (B) 160 cm   (C) 320 cm   (D) 640 cm   (E) 1280 cm

Problems from Year 2001

Problems 3 points each

1. Julia has four candles. Each of her candles burns itself out within three hours. Julia lit two candles and placed them next to an open window. 30 minutes later the wind blew out one candle, and an hour later the wind blew out the second candle. Then Julia closed the window and lit all four candles. How long after this moment will the last candle burn out?

    (A) 1 hr 30 min   (B) 2 hr   (C) 3 hr   (D) 7 hr 30 min   (E) 8 hr

2. Joseph had 7 sticks. He broke one of them into two pieces. How many sticks does Joseph have now?

    (A) 5   (B) 6   (C) 7   (D) 8   (E) 9

3. Kuba bought a chocolate heart for his mother (see the picture). Each chocolate square weighs 10 grams. What is the weight of the whole heart?

    (A) 340 grams   (B) 360 grams   (C) 380 grams
    (D) 400 grams   (E) 420 grams

4. There are 12 pairs of shoes on each of 10 shelves in the animal shoe store. 5 centipedes are the first customers. Three of them buy 30 pairs of shoes each, and two of them buy 5 pairs of shoes each. How many pairs of shoes are left on the shelves?

    (A) 10   (B) 15   (C) 20   (D) 25   (E) 30

5. For five days, Kaya was helping her mother pick berries. On the first day, she ate most of her berries and gave her mother only one cup of berries. She decided that each day she would be giving her mother twice as much berries as the day before. How many cups of berries did Kaya give her mother over five days?

    (A) 5   (B) 31   (C) 21   (D) 11   (E) 16

6. Which of the following expressions is correct?

   (A) 12 ÷ (4 + 8) = 11   (B) 8 × 2 + 3 = 40   (C) 2 × 3 + 4 × 5 = 50
   (D) (10 + 8) ÷ 2 = 14   (E) 18 − 6 ÷ 3 = 16

7. There are 19 girls and 12 boys in the school yard. At least how many students need to join them so that six groups with the same number of students can be formed?

   (A) 1   (B) 2   (C) 3   (D) 4   (E) 5

8. Four sticks, each 14 cm long, were placed in the way shown in the picture, for a total length of 80 cm. The distances between the sticks are equal. How long is each of these distances?

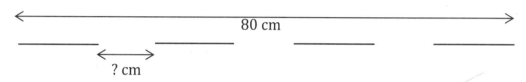

   (A) 1 cm   (B) 2 cm   (C) 3 cm   (D) 5 cm   (E) 8 cm

Problems 4 points each

9. Bobby was born on Abby's third birthday. How many years later will Abby be twice as old as Bobby?

   (A) 1   (B) 2   (C) 3   (D) 4   (E) 5

10. The lowest point of the Sniezna Cave in the Tatra Mountains is located 221 m below the cave entrance, and the highest point of the Sniezna Cave is located 419 m above the cave entrance. What is the depth (the distance between the lowest and highest points) of this cave?

   (A) 198 m   (B) 221 m   (C) 419 m   (D) 640 m   (E) 650 m

11. I divided 20 pieces of candy among several children. Each child received at least one piece of candy, and everyone received a different number of pieces of candy. What is the greatest possible number of children who received candy?

   (A) 20   (B) 10   (C) 8   (D) 6   (E) 5

12. Which figure is different from all the others?

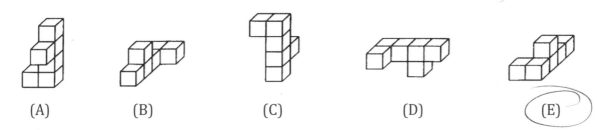

(A) (B) (C) (D) (E)

13. Ella and Bella got on a super-train. Ella took a seat in the seventeenth car counting from the front of the train, and Bella was seated in the thirty-fourth car counting from the end. The girls were sitting in the same car. How many cars did the super-train have?

(A) 48   (B) 49   (C) 50   (D) 51   (E) 52

14. Adam and Bart are picking up chestnuts. At a certain moment they have the same number of chestnuts, and then Adam gives Bart half of all his chestnuts. How many times is the number of chestnuts Bart has greater than the number of chestnuts Adam has now?

(A) 2 times   (B) 3 times   (C) 4 times   (D) 5 times
(E) It depends on the number of chestnuts they had at the beginning.

15. There are squares and triangles on the table. They have 17 vertices altogether. How many triangles are there on the table?

(A) 1   (B) 2   (C) 3   (D) 4   (E) 5

16. What is the least number of matches that must be added to the picture in order to get exactly 11 squares?

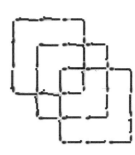

(A) 2   (B) 3   (C) 4   (D) 5   (E) 6

Problems 5 points each

17. In a certain picture you can see numbers 1, 2, 3, 4, 5 with their mirror reflections.

What is the next picture in the sequence of reflections?

(A)   (B)   (C)   (D)   (E)

18. Which of the five napkins below comes from this paper cut-out?

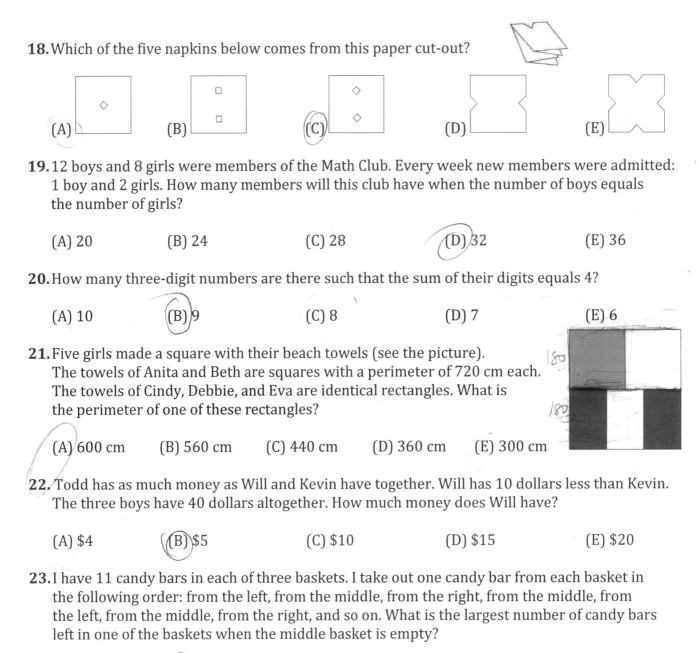

19. 12 boys and 8 girls were members of the Math Club. Every week new members were admitted: 1 boy and 2 girls. How many members will this club have when the number of boys equals the number of girls?

(A) 20 (B) 24 (C) 28 (D) 32 (E) 36

20. How many three-digit numbers are there such that the sum of their digits equals 4?

(A) 10 (B) 9 (C) 8 (D) 7 (E) 6

21. Five girls made a square with their beach towels (see the picture). The towels of Anita and Beth are squares with a perimeter of 720 cm each. The towels of Cindy, Debbie, and Eva are identical rectangles. What is the perimeter of one of these rectangles?

(A) 600 cm (B) 560 cm (C) 440 cm (D) 360 cm (E) 300 cm

22. Todd has as much money as Will and Kevin have together. Will has 10 dollars less than Kevin. The three boys have 40 dollars altogether. How much money does Will have?

(A) $4 (B) $5 (C) $10 (D) $15 (E) $20

23. I have 11 candy bars in each of three baskets. I take out one candy bar from each basket in the following order: from the left, from the middle, from the right, from the middle, from the left, from the middle, from the right, and so on. What is the largest number of candy bars left in one of the baskets when the middle basket is empty?

(A) 1 (B) 2 (C) 5 (D) 6 (E) 11

24. The sum of the dots on the opposite sides of a die is seven. We move a die on a grid as the picture shows. The die is rotated once for each square as it is moved as shown by the arrows. How many dots are on the top of the die when it is located on the square marked with *?

(A) 5   (B) 4   (C) 3   (D) 1
(E) other number

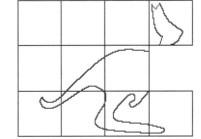

Problems from Year 2002

Problems 3 points each

1. Which of the squares below should be put into the picture to the right to get the symbol of our competition?

(A)   (B)   (C)   (D)   (E)

2. After we simplify 2 + 2 − 2 + 2 − 2 + 2 − 2 + 2 − 2 + 2 the result will be:

(A) 0   (B) 2   (C) 4   (D) 12   (E) 20

3. Andrzej received three cars, four balls, three teddy bears, ten pens, two chocolate bars, and a book for his birthday. How many items did he get in all?

(A) 15   (B) 17   (C) 20   (D) 23   (E) 27

4. A square was divided into pieces (see the picture). Which of the pieces below does not occur in this divided square?

(A)   (B)   (C)   (D)   (E)

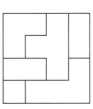

5. Gina, Lena, Suzie, and Joanna have their birthdays on March 1st, May 17th, July 20th, and March 20th (not necessarily in that order). Lena and Suzie were born in the same month. Gina and Suzie were born on the same day of the month. Which of the girls was born on May 17th?

(A) Gina   (B) Lena   (C) Suzie   (D) Joanna
(E) This cannot be determined from the information given.

PROBLEMS 2002

6. The human heart beats an average of 70 times per minute. On average how many times does it beat during one hour?

   (A) 42,000   (B) 7,000   (C) 4,200   (D) 700   (E) 420

7. Quadrilateral ABCD is a square and its side is 10 cm long. Quadrilateral ATMD is a rectangle and its shorter side is 3 cm. What is the difference between the sum of the lengths of all the sides of the square and the sum of the lengths of all the sides of the rectangle ATMD?

   (A) 14 cm   (B) 10 cm   (C) 7 cm   (D) 6 cm   (E) 4 cm

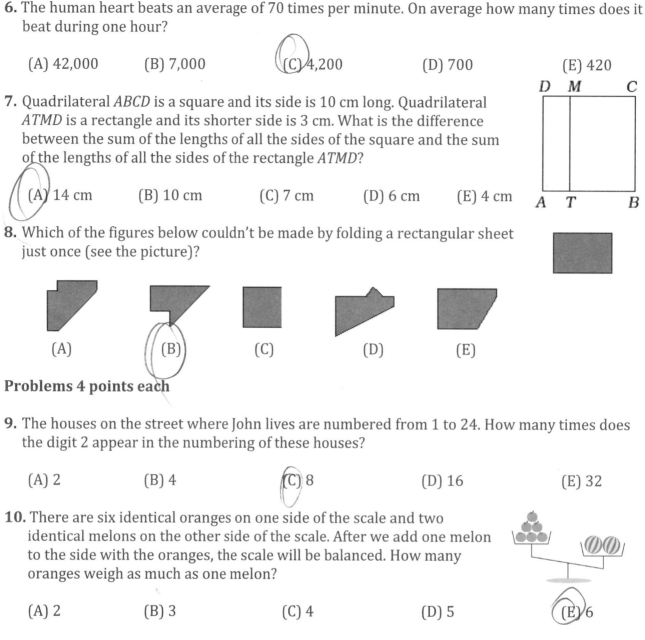

8. Which of the figures below couldn't be made by folding a rectangular sheet just once (see the picture)?

   (A)   (B)   (C)   (D)   (E)

Problems 4 points each

9. The houses on the street where John lives are numbered from 1 to 24. How many times does the digit 2 appear in the numbering of these houses?

   (A) 2   (B) 4   (C) 8   (D) 16   (E) 32

10. There are six identical oranges on one side of the scale and two identical melons on the other side of the scale. After we add one melon to the side with the oranges, the scale will be balanced. How many oranges weigh as much as one melon?

    (A) 2   (B) 3   (C) 4   (D) 5   (E) 6

11. This picture below is a sketch of a castle. Which of the lines below is not part the sketch?

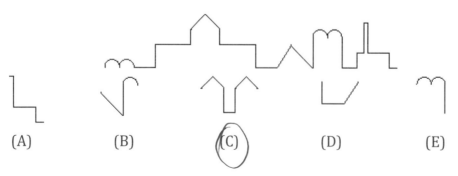

    (A)   (B)   (C)   (D)   (E)

PROBLEMS 2002

12. We add 17 to the smallest two-digit number and then we divide the sum by the largest one-digit number. What is the result?

(A) 3 (B) 6 (C) 9 (D) 11 (E) 27

13. In a certain ancient country the numbers one, ten, and sixty were expressed with the following symbols:

one ten sixty

People were writing down other numbers using these symbols. For example, the number 22 was written as:

Which of the following notations represents the number 124?

(A)    (B)    (C)    (D)    (E)

14. The face of a clock was divided into four parts. The sums of the numbers in each of those parts are consecutive numbers. Which of the following pictures satisfies this rule?

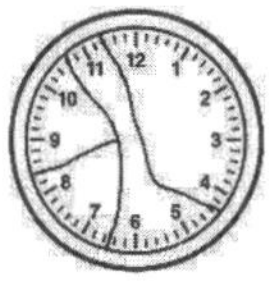

(A) 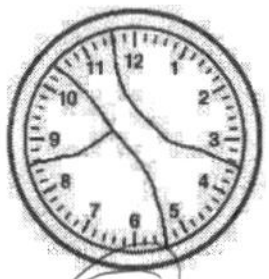 (B)  (C)  (D) 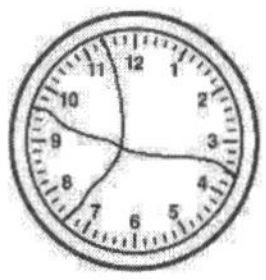 (E)

15. Klara and Zoe had 60 matches altogether. Klara took as many matches as she needed to build a triangle with each side 6 matches long. Zoe used the remaining matches to build a rectangle which had one side equal to 6 matches. How many matches did she use to make one longer side of this rectangle?

(A) 9 (B) 12 (C) 15 (D) 18 (E) 30

16. Three kangaroos, Miki, Niki, and Oki, participated in a competition. Jumping at the same speed, they jumped along the lines you can see in the picture. Only one of the sentences A, B, C, D, and E listed below is true. Which one is it?

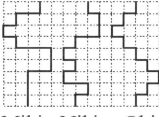

Miki   Niki   Oki

(A) Miki and Oki finished at the same time.
(B) Niki finished first.
(C) Oki finished last.
(D) All the kangaroos finished at the same time.
(E) Miki and Niki finished at the same time.

Problems 5 points each

17. Each boy, Mike, Nate, Oliver, and Paul, has exactly one of the following animals: a cat, a dog, a goldfish, and a canary. Nate has a pet with fur. Oliver has a pet with four legs. Paul has a bird, and Mike and Nate don't like cats. Which of the following sentences is not true?

    (A) Oliver has a dog.   (B) Paul has a canary.   (C) Mike has a goldfish.
    (D) Oliver has a cat.   (E) Nate has a dog.

18. Mary leaves her house at 6:55 and arrives at school at 7:32. Zoe needs 12 minutes less than Mary to get to school. Yesterday Zoe showed up at school at 7:45. What time did she leave her house?

    (A) at 7:07   (B) at 7:20   (C) at 7:25   (D) at 7:30   (E) at 7:33

19. Robert had a certain number of identical cubes. He made a tunnel using half of his blocks (see Picture 1). With some of the remaining cubes he formed a pyramid (see Picture 2). How many blocks were left after making both of these structures?

    (A) 34   (B) 28   (C) 22   (D) 18   (E) 15

    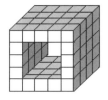 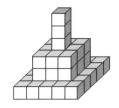
    Picture 1   Picture 2

20. The daughter is 3 years old, and her mom is 28 years older than the daughter. How many years later will the mom be three times as old as her daughter?

    (A) 9   (B) 12   (C) 10   (D) 1   (E) 11

21. A conductor wanted to make a trio consisting of a violinist, a pianist, and a drummer. He had to choose one of two violinists, one of two pianists, and one of two drummers. He decided to try each of the possible trios. How many attempts did he have to make?

   (A) 3   (B) 4   (C) 8   (D) 24   (E) 25

22. One medal can be cut out from a golden square plate. If four medals are made from four plates, the remaining parts of those four plates can be used to make one more plate. What is the largest number of medals that can be made when 16 plates are used?

   (A) 17   (B) 19   (C) 20   (D) 21   (E) 32

23. Twenty eight students from the fourth grade competed in a math competition. Each student earned a different number of points. The number of students who received more points than Tom is two times smaller than the number of students who had fewer points than Tom. In which position did Tom finish that competition?

   (A) 6th   (B) 7th   (C) 8th   (D) 9th   (E) 10th

24. An odometer in a car shows the number 187569, which is the number of kilometers that have been traveled. This number consists of all different digits. After traveling how many more kilometers will the odometer show a number consisting of all different digits again?

   (A) after 777 km   (B) after 12,431 km   (C) after 431 km   (D) after 21 km   (E) after 11 km

Problems from Year 2003

Problems 3 points each

1. The picture shows the letter U drawn on grid paper. How many squares does the letter U cover?

   (A) 10   (B) 8   (C) 11   (D) 13   (E) 12

2. What is the result of 0 + 1 + 2 + 3 + 4 − 3 − 2 − 1 − 0?

   (A) 0   (B) 2   (C) 4   (D) 10   (E) 16

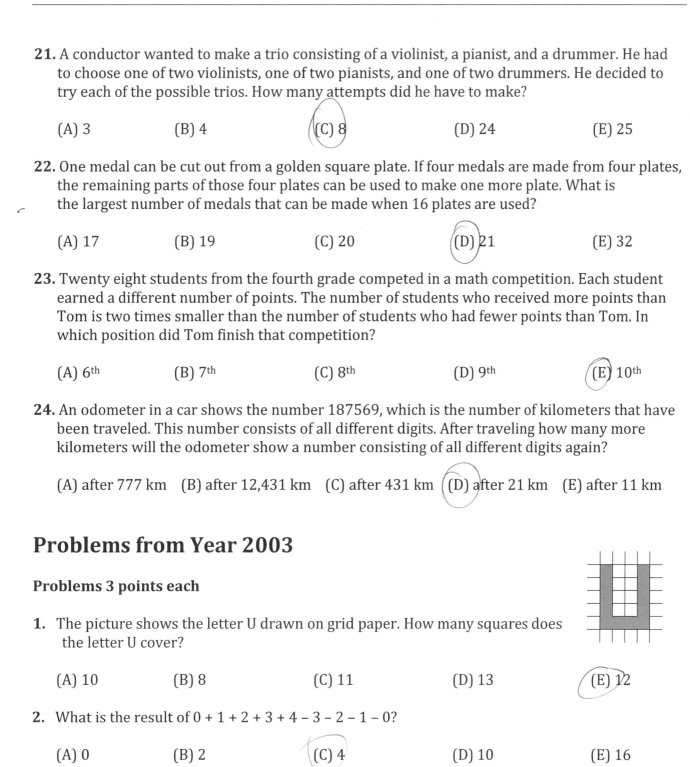

3. The first car of the train, right behind the engine, contains 10 boxes. In each of the following cars there are twice as many boxes as in the car in front of it. How many boxes are there in the fifth car?

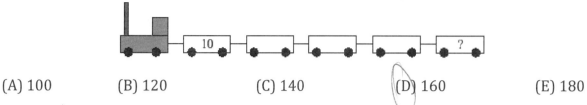

(A) 100   (B) 120   (C) 140   (D) 160   (E) 180

4. Sophia is drawing kangaroos. The first one is blue, the next one green, the one after it red, the fourth one yellow, and then again blue, green, red, yellow, and so on, in the same order. What color will the seventeenth kangaroo be?

(A) blue   (B) green   (C) red   (D) black   (E) yellow

5. In the teachers' lounge there are 6 tables with 4 chairs each, 4 tables with 2 chairs each, and 3 tables with 6 chairs each. How many chairs are there in the lounge?

(A) 40   (B) 25   (C) 50   (D) 36   (E) 44

6. In one of these pictures, there are three times as many hearts as other shapes. Which picture is it?

7. A rectangle with dimensions 7 × 4 was outlined on grid paper. How many squares will a diagonal of this rectangle intersect?

(A) 8   (B) 9   (C) 10   (D) 11   (E) 12

8. The figure presented in the picture, which is made out of a certain number of identical cubes, weighs 189 grams. How much does one cube weigh?

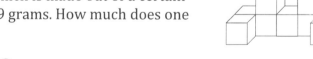

(A) 29 grams   (B) 25 grams   (C) 21 grams   (D) 19 grams   (E) 17 grams

Problems 4 points each

9. Philip wrote down consecutive natural numbers starting with 3 until he had written 35 digits. What was the greatest number that Philip wrote down?

(A) 12   (B) 22   (C) 23   (D) 28   (E) 35

PROBLEMS 2003

10. Anna fell asleep at 9:30 p.m. and woke up at 6:45 a.m. the next day. Her little brother Peter slept 1 hour and 50 minutes longer. How long did Peter sleep?

(A) 30 hr 5 min (B) 11 hr 35 min (C) 11 hr 5 min (D) 9 hr 5 min (E) 8 hr 35 min

11. A pattern, the beginning and the end of which is shown in the picture, is made up of alternating black and white bars. There are 17 bars altogether. The bars on both ends are black. How many white bars are there in the pattern?

(A) 9 (B) 16 (C) 7 (D) 8 (E) 15

12. Jumping Kangaroo is practicing for the animal Olympics. His longest jump during the training was 55 dm 50 mm long, but in the finals of the Olympics he won with a jump that was 123 cm longer. How long was Jumping Kangaroo's longest jump during the Olympics? (Remember that 1 m = 10 dm, 1 dm = 10 cm, and 1 cm = 10 mm)

(A) 6 m 78 cm (B) 5 m 73 cm (C) 5 m 55 cm (D) 11 m 28 cm (E) 7 m 23 cm

13. Paul chose a certain number, subtracted 203 from it, and then added 2003 to the difference. His final result was 20003. What number did Paul choose at the beginning?

(A) 23 (B) 17797 (C) 18203 (D) 21803 (E) 22209

14. Barbara likes to add the digits showing the current time on her electronic watch (for example, when the watch shows 21:17, she gets the number 11 as the result). What is the greatest sum she can get?
(Hint: In some countries and sometimes in US, instead of saying "It is 1 p.m.," people may say, "It is 13:00." When it is 2 p.m., they may say, "It is 14:00," and when it is 12 a.m., they may say, "It is 24:00." In this problem, 21:17 means 9:17 p.m. Time expressed in this way is sometimes called "military time.")

(A) 24 (B) 36 (C) 19 (D) 25 (E) 23

15. Mark said to his friends, "If I had picked twice as many apples as I did, I would have 24 more apples than I have now." How many apples did Mark pick?

(A) 48 (B) 24 (C) 42 (D) 12 (E) 36

16. Points A, B, C, and D, all of which lie on a straight line, are marked in the picture below. The distance between points A and C is 10 m, between B and D is 15 m, and between A and D is 22 m. What is the distance between points B and C?

(A) 1 m (B) 2 m (C) 3 m (D) 4 m (E) 5 m

Problems 5 points each

17. There are 29 students in a certain class. 12 of the students have a sister and 18 of the students have a brother. In this class, only Tania, Barbara, and Anna do not have any siblings. How many students from this class have both a brother and a sister?

(A) None   (B) 1   (C) 3   (D) 4   (E) 6

18. Daniel has 11 pieces of paper. He cut some of them into three parts and now he has 29 pieces of paper. How many pieces of paper did he cut?

(A) 3   (B) 2   (C) 8   (D) 11   (E) 9

19. John bought 3 kinds of cookies: large, medium, and small. The large cookies cost 4 dollars each, the medium 2 dollars each, and the small 1 dollar each. Altogether, John bought 10 cookies and paid 16 dollars. How many large cookies did he buy?

(A) 5   (B) 4   (C) 3   (D) 2   (E) 1

20. Christopher built the rectangular prism shown in the picture using red and blue cubes of identical size. The outer walls of this prism are red but all the inner cubes are blue. How many blue cubes did Christopher use in this construction?

(A) 12   (B) 16   (C) 22   (D) 26   (E) 32

21. Jerry is planning to buy some basketballs. If he buys 5 basketballs, he will have 10 dollars left over, and if he buys 7 basketballs, he will have to borrow 22 dollars. How many dollars does one basketball cost?

(A) 11   (B) 16   (C) 22   (D) 26   (E) 32

22. Mark built a rectangular prism using 3 blocks, each of which is made out of 4 small cubes connected in various ways. Two of the blocks are shown in the picture. Which is the third (white) block with only two visible sides?

23. Two pieces that made up the shaded region were cut out from a square puzzle (see the picture). Which two of the pieces are they?

1   2   3   4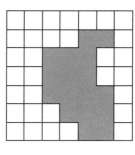

(A) 1 and 3   (B) 2 and 4   (C) 2 and 3   (D) 1 and 4   (E) 3 and 4

24. At the toy store, among other things, you can buy dogs, bears, and kangaroos. Three dogs and two bears together cost as much as four kangaroos. For the same amount of money, you can buy one dog and three bears. Which of the statements below is true?

(A) A dog is twice as expensive as a bear.
(B) A bear is twice as expensive as a dog.
(C) The prices of a dog and of a bear are identical.
(D) A bear is three times as expensive as a dog.
(E) A dog is three times as expensive as a bear.

Problems from Year 2004

Problems 3 points each

1. 2001 + 2002 + 2003 + 2004 + 2005 =

   (A) 1,015   (B) 5,010   (C) 10,150   (D) 11,005   (E) 10,015

2. Mark was 4 years old when his sister was born. Today he blew out all 9 candles on his birthday cake. What is the difference between Mark's age and his sister's age today?

   (A) 4 years   (B) 5 years   (C) 9 years   (D) 13 years   (E) 14 years

3. The picture below shows a road from town A to town B (indicated by a solid line) and a detour (marked by a dashed line) caused by renovations of the section CD. How many kilometers longer is the road from town A to town B because of the detour?

   (A) 3 km   (B) 5 km   (C) 6 km   (D) 10 km   (E) This cannot be calculated.

4. Which of the results below is not identical to the difference 671 – 389?

   (A) 771 – 489   (B) 681 – 399   (C) 669 – 391   (D) 1871 – 1589   (E) 600 – 318

5. There were some birds sitting on the telegraph wire. Then, 5 of them flew away and after some time 3 birds came back. At that time there were 12 birds sitting on the wire. How many birds were there at the very beginning?

   (A) 8   (B) 9   (C) 10   (D) 12   (E) 14

6. Which numbers are inside the rectangle and inside the circle but not inside the triangle as well?

   (A) 5 and 11   (B) 1 and 10   (C) 13
   (D) 3 and 9   (E) 6, 7 and 4

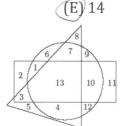

7. The buildings on Color Street are numbered from 1 to 5 (see the picture). Each building is colored with one of the following colors: blue, red, yellow, pink, and green. We know that:
   – The red building is next only to the blue building.
   – The blue building is between the red building and the green building.
   What is the color of the building with the number 3?

   (A) blue   (B) red   (C) yellow   (D) pink   (E) green

8. How many white squares need to be shaded in the picture to the right so that the number of shaded squares is equal to exactly half of the number of white squares?

   (A) 2   (B) 3   (C) 4   (D) 6   (E) It is impossible to calculate.

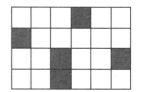

Problems 4 points each

9. Five identical rectangular plastic sheets were divided into white and black squares. Which of the sheets from A to E has to be covered with the sheet to the right in order to get a completely black rectangle?

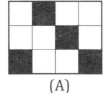 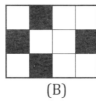 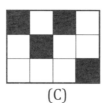 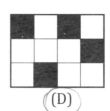 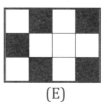

   (A)   (B)   (C)   (D)   (E)

10. The scales in the pictures have been balanced. There are pencils and a pen on the two sets of scales. What is the weight of the pen in grams?

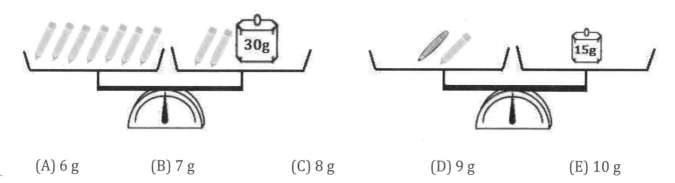

(A) 6 g   (B) 7 g   (C) 8 g   (D) 9 g   (E) 10 g

11. I noticed four clocks on the wall (see the picture). Only one of them shows the correct time. One of them is 20 minutes ahead, another is 20 minutes late, and yet another one is broken. What time is it right now?

(A) 4:45   (B) 5:05   (C) 5:25
(D) 5:40   (E) 12:00

12. Ella brought a basket of apples and oranges to a birthday party. Guests ate half of the apples and a third of the oranges. What remained in the basket was:

(A) half of all the fruit.   (B) more than half of all the fruit.   (C) less than half of all the fruit.
(D) a third of all the fruit.   (E) less than a third of all the fruit.

13. Angie divided a certain number by 10 instead of multiplying it by 10. As a result she got 600. What would the result have been if she hadn't made this mistake?

(A) 6   (B) 60   (C) 600   (D) 6,000   (E) 60,000

14. Kathy found a book which was missing some pages. When she opened the book, she saw the number 24 on the left side and the number 45 on the right side. How many sheets of paper were missing between these two pages?

(A) 9   (B) 10   (C) 11   (D) 20   (E) 21

15. Eva is 52 days older than her friend Andrea. Eva had her birthday on Tuesday in March this year. On which day of the week will Andrea celebrate her birthday this year?

(A) Monday   (B) Tuesday   (C) Wednesday   (D) Thursday   (E) Friday

16. Numbers were placed in a square table ⊞ in such a way that the sum of the numbers in the first row is equal to 3, the sum of the numbers in the second row is equal to 8, and the sum of the numbers in the first column is equal to 4. What is the sum of the numbers in the second column?

(A) 4   (B) 6   (C) 7   (D) 8   (E) 11

Problems 5 points each

17. A certain cube is painted with three colors so that every side of this cube is one of the colors and opposite sides are the same color. From which of the patterns below can this kind of cube be made?

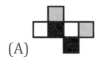 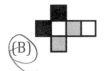 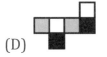
(A)   (B)   (C)   (D)   (E)

18. Four square tiles were arranged in the way shown in the picture. The lengths of the sides of two tiles are indicated. What is the length of the side of the largest tile?

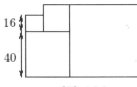

(A) 24   (B) 56   (C) 64   (D) 81   (E) 100

19. Girls and boys from Maria and Matt's class formed a line. There are 16 students on Maria's right, and Matt is among them. There are 14 students on Matt's left, and Maria is among them. There are 7 students between Maria and Matt. How many students are there in this class?

(A) 37   (B) 30   (C) 23   (D) 22   (E) 16

20. The sum of the digits of the 10-digit number is 9. What is the product of the digits of this number?

(A) 0   (B) 1   (C) 45   (D) 9 × 8 × 7 × ... × 2 × 1   (E) 10

21. The big cube was formed out of 125 small cubes, each of which is either black or white (see the picture). Any two adjacent cubes have different colors. The vertices of the big cube are black. How many white cubes does the big cube contain?

(A) 62   (B) 63   (C) 64   (D) 65   (E) 68

PROBLEMS 2004

22. A lottery ticket costs 4 dollars. Three boys, Paul, Peter, and Robert, collected money and bought two tickets. Paul gave 1 dollar, Peter gave 3 dollars, and Robert gave 4 dollars. One of the tickets they bought won 1000 dollars. Boys shared the award fairly, that is, proportionally to their contributions. How much did Peter receive?

(A) $300 (B) $375 (C) $250 (D) $750 (E) $425

23. In three soccer games Daniel's team scored three goals and had one goal scored against them. For every game won the team gets 3 points, for a tie it gets 1 point, and for a game lost it gets 0 points. It is certain that the number of points the team earned in those three games was **not equal** to which of the following numbers?

(A) 7 (B) 6 (C) 5 (D) 4 (E) 3

24. Natural numbers were placed in a table. In each white cell of the table, the products of two numbers from the gray sections—one from above and one from the left—was placed (for example: 42 = 7 × 6). Some of these products are represented by letters. Which two letters represent the same number?

(A) L and M (B) T and N (C) R and P (D) K and P (E) M and S

Problems from Year 2005

Problems 3 points each

1. A butterfly sat down on a correctly solved problem. What number did it cover up?

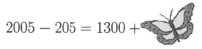

(A) 250 (B) 400 (C) 500 (D) 910 (E) 1800

2. At noon, the minute hand of a clock is in the position shown in the picture on the right. What will the position of the minute hand be after 17 quarters of an hour pass?

(A) (B) (C) (D) (E)

© Math Kangaroo in USA, NFP

3. Joan bought some cookies, each of which costs 3 dollars. She gave the salesperson 10 dollars and received 1 dollar as change. How many cookies did Joan buy?

   (A) 2   (B) 3   (C) 4   (D) 5   (E) 6

4. After the trainer's first whistle, the monkeys at the circus formed 4 rows. There were 4 monkeys in each row. After the second whistle, they rearranged themselves into 8 rows. How many monkeys were there in each row after the second whistle?

   (A) 1   (B) 2   (C) 3   (D) 4   (E) 5

5. Eva lives with her parents, her brother, one dog, two cats, two parrots, and four fish. What is the total number of legs that they have altogether?

   (A) 22   (B) 24   (C) 28   (D) 32   (E) 40

6. John has a chocolate bar made up of square pieces 1 cm × 1 cm in size. He has already eaten some of the corner pieces (see the picture). How many pieces does John have left?

   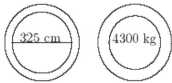

   (A) 66   (B) 64   (C) 62   (D) 60   (E) 58

7. Two traffic signs mark the bridge in my village (see the picture below). These signs indicate the maximum vehicle width and the maximum vehicle weight allowed on the bridge. Which one of the following trucks is allowed to cross that bridge?

   (A) a truck that is 315 cm wide and weighs 4400 kg
   (B) a truck that is 330 cm wide and weighs 4250 kg
   (C) a truck that is 325 cm wide and weighs 4400 kg
   (D) a truck that is 330 cm wide and weighs 4200 kg
   (E) a truck that is 325 cm wide and weighs 4250 kg

8. Each of seven boys paid the same amount of money for a trip. The total sum of what they paid is a three-digit number which can be written as 3☐0. What is the middle digit of this number?

   (A) 3   (B) 4   (C) 5   (D) 6   (E) 7

Problems 4 points each

9. What is the smallest possible number of children in a family where each child has at least one brother and at least one sister?

(A) 2   (B) 3   (C) 4   (D) 5   (E) 6

10. From the five numbers below, I chose one number. The number is even and all of its digits are different. The hundreds digit is double the ones digit. The tens digit is greater than the thousands digit. Which number did I choose?

(A) 1246   (B) 3874   (C) 4683   (D) 4874   (E) 8462

11. A square piece of paper has been cut into three pieces. Two of them are shown in the picture on the right. Which of the pieces below is the third one?

(A)   (B)   (C)   (D)   (E)

12. An elevator cannot carry more than 150 kg. Four friends weigh 60 kg, 80 kg, 80 kg, and 80 kg, respectively. What is the least number of trips necessary to carry the four friends to the highest floor?

(A) 1   (B) 2   (C) 3   (D) 4   (E) 7

13. Ala has 24 dollars and Barb has 66 dollars. Sophia has as much less than Barb as she has more than Ala. How many dollars does Sophia have?

(A) 33   (B) 35   (C) 42   (D) 45   (E) 48

14. There are eight kangaroos in the cells of the table (see the picture). What is the least number of kangaroos that need to be moved to the empty cells so that there would be exactly two kangaroos in any row and in any column of the table?

(A) 4   (B) 3   (C) 2   (D) 1   (E) 0

15. Greg has a sack with a hole. He needs to bring four full sacks of sand from the river to a house located at the other end of the village. Unfortunately, every time he goes through the village, half of the sand spills out of the sack through the hole. How many trips does Greg need to make from the river to the house in order to bring the required amount of sand?

(A) 4   (B) 5   (C) 6   (D) 7   (E) 8

16. During a Kangaroo camp, Adam solved five problems every day and Brad solved two problems daily. After how many days did Brad solve as many problems as Adam solved in 4 days?

(A) after 5 days (B) after 7 days (C) after 8 days (D) after 10 days (E) after 20 days

Problems 5 points each

17. There were 9 pieces of paper. Some of them were cut into three pieces. As a result, there are now 15 pieces of paper now. How many pieces of paper were cut?

(A) 2 (B) 3 (C) 4 (D) 5 (E) 6

18. Using 6 matches, only one rectangle with a perimeter of 6 matches can be made (see the picture). How many different rectangles with a perimeter of 14 matches can be made using 14 matches?

(A) 2 (B) 3 (C) 4 (D) 6 (E) 12

19. A picture frame was constructed using pieces of wood which all have the same width. What is the width of the frame if the inside perimeter of the frame is 8 decimeters (dm) less than its outside perimeter?

(A) 1 dm (B) 2 dm (C) 4 dm (D) 8 dm
(E) It depends on the size of the picture.

20. In a certain trunk there are 5 chests, in each chest there are 3 boxes, and in each box there are 10 gold coins. The trunk, the chests, and the boxes are locked. At least how many locks need to be opened in order to take out 50 coins?

(A) 5 (B) 6 (C) 7 (D) 8 (E) 9

21. The figure shows a rectangular garden with dimensions of 16 m by 20 m. The gardener planted six identical flowerbeds (colored gray in the diagram). What is the perimeter of each of the flowerbeds?

(A) 20 m (B) 22 m (C) 24 m (D) 26 m (E) 28 m

22. Mike chose a three-digit number and a two-digit number. The difference of these numbers is 989. What is their sum?

(A) 1001 (B) 1010 (C) 2005 (D) 1000 (E) 1009

23. Five cards are lying on the table in the order 5, 1, 4, 3, 2, as shown in the top row of the picture. They need to be placed in the order shown in the bottom row. In each move, any two cards may be switched. What is the least number of moves that need to be made?

(A) 2   (B) 3   (C) 4   (D) 5   (E) 6

24. Which of the cubes has the plan shown in the picture to the right?

(A)   (B)   (C)   (D)   (E)

Problems from Year 2006

Problems 3 points each

1. During a summer math camp in the city of Zakopane in Poland, the students take part in a trip to the top of Mount Giewont. It takes 3 hours to get to the top of the mountain. They stay at the top of the mountain for half an hour. Afterwards, it takes two and a half hours to come down the mountain. What time in the morning at the latest does the trip need to start so that everybody is back at the camp for a meal at 3 p.m.?

   (A) 8:00   (B) 8:30   (C) 9:00   (D) 9:30   (E) 10:00

2. What is the value of this expression: $2 \times 0 \times 0 \times 6 + 2006$?

   (A) 0   (B) 2006   (C) 2014   (D) 2018   (E) 4012

3. How many cubes have been removed from the first block to obtain the second block?

   (A) 4   (B) 5   (C) 6   (D) 7   (E) 9

4. Katie's birthday was yesterday. It will be Thursday tomorrow. What day was Katie's birthday?

   (A) Tuesday   (B) Wednesday   (C) Thursday   (D) Saturday   (E) Monday

5. John is playing with darts. He brings back all the darts after he has thrown them, and for each time he hits the bullseye, he gains two additional darts. At the beginning he has 10 darts and at the end he has 20. How many times did he hit the bullseye?

   (A) 6   (B) 8   (C) 10   (D) 5   (E) 4

6. A kangaroo enters the building as shown in the picture. He only passes through triangular rooms. Where does he leave the building?

   (A) a   (B) b   (C) c   (D) d   (E) e

7. Four people can sit around a certain kind of a square table. For the school party the students put together 7 such square tables in order to make one long rectangular table. How many people can sit at this long table now?

   (A) 14   (B) 16   (C) 21   (D) 24   (E) 28

8. In his wallet, Stan has one 5-dollar bill, one 2-dollar bill, and one 1-dollar bill. Which of the following amounts can Stan not make using the bills that he has?

   (A) $3   (B) $4   (C) $6   (D) $7   (E) $8

Problems 4 points each

9. On one side of Long Street the houses are numbered with consecutive odd numbers from 1 to 19. On the other side of that street, the houses are numbered with consecutive even numbers from 2 to 14. How many houses are there on Long Street?

   (A) 8   (B) 16   (C) 17   (D) 18   (E) 33

10. From which of the figures below was the figure to the right cut out?

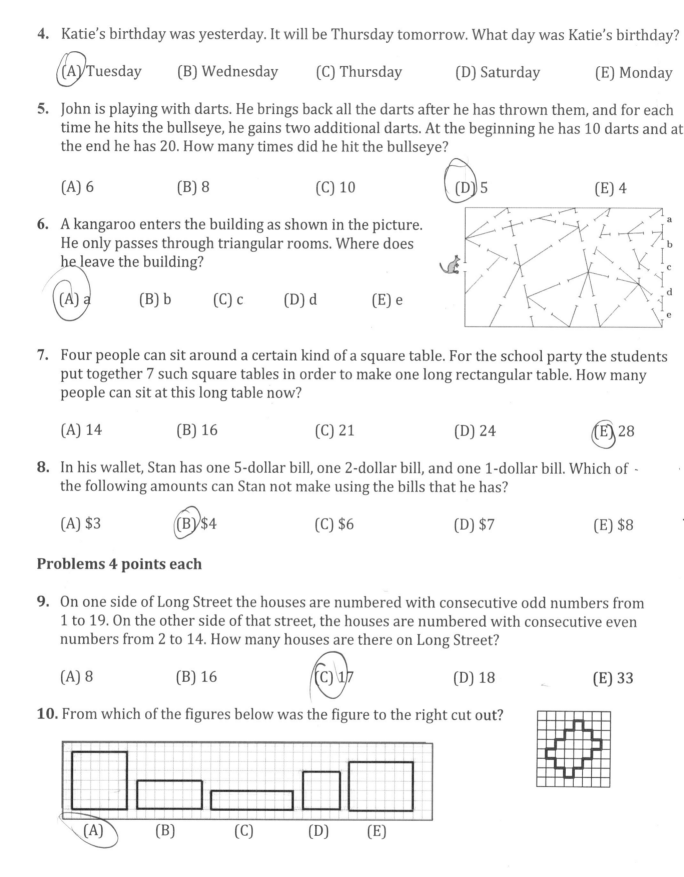

PROBLEMS 2006

11. The picture below shows bus routes and ticket prices between 6 towns. What is the least amount of money needed to pay for the tickets to get from town A to town B?

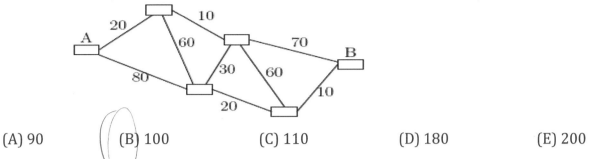

(A) 90   (B) 100   (C) 110   (D) 180   (E) 200

12. What is the smallest number we can get by arranging the six cards with numbers shown in the picture in one row, one after another?

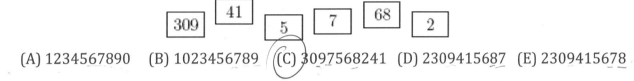

(A) 1234567890   (B) 1023456789   (C) 3097568241   (D) 2309415687   (E) 2309415678

13. Six weights, weighing 1 pound, 2 pounds, 3 pounds, 4 pounds, 5 pounds, and 6 pounds, were placed into three boxes, two weights in each box. The weights in the first box weigh 9 pounds together, and those in the second box weigh 8 pounds. Which weights are in the third box?

(A) 5 and 2   (B) 6 and 1   (C) 3 and 1   (D) 4 and 2   (E) 4 and 3

14. Four routes are drawn between two points. Which route is the shortest?

(A)    (B)   (C)   (D)

(E) All are equal.

15. Numbers are written on a "number flower." Mary picked off all the petals with numbers which give a remainder 2 when divided by 6. What is the sum of the numbers on the petals that Mary picked off?

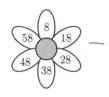

(A) 46   (B) 66   (C) 84   (D) 86   (E) 114

16. You can move or rotate any of the shapes of the puzzle, but you cannot flip them over. Which of the shapes below does not appear in the puzzle to the right?

(A)   (B)   (C)   (D)   (E)

Problems 5 points each

17. Four crows are sitting on a fence. Their names are Dana, Hanna, Lena, and Bennie. Dana sits exactly in the middle between Hanna and Lena. The distance between Hanna and Dana is the same as the distance between Lena and Bennie. Dana sits 4 feet away from Bennie. How far away is Hanna sitting from Bennie?

(A) 5 feet   (B) 6 feet   (C) 7 feet   (D) 8 feet   (E) 9 feet

18. Johnny is building a house out of cards. In the picture, one-story, two-story, and three-story houses are shown. How many cards does Johnny need to build a 4-story house?

(A) 23   (B) 24   (C) 25   (D) 26   (E) 27

19. The structure shown in the picture is made by gluing together the sides of 10 cubes. Roman painted the entire structure, including the bottom. How many faces of the cubes did he paint?

(A) 18   (B) 24   (C) 30   (D) 36   (E) 42

20. Irena, Ann, Kate, Olga, and Elena live in the same two-story house. Two of the girls live on the first floor; three of them live on the second floor. Olga lives on a different floor than Kate and Elena. Ann lives on a different floor than Irena and Kate. Who lives on the first floor?

(A) Kate and Elena   (B) Irena and Elena   (C) Irena and Olga
(D) Irena and Kate   (E) Ann and Olga

21. In the expression 2002 □ 2003 □ 2004 □ 2005 □ 2006, either "+" or "–" can be written in the place of each □. Which result is impossible?

(A) 1998   (B) 2001   (C) 2002   (D) 2004   (E) 2006

22. One year in March, there were 5 Mondays. Which day of the week below could not appear in this month five times as well?

(A) Saturday   (B) Sunday   (C) Tuesday   (D) Wednesday   (E) Thursday

23. In each of the nine cells of the square, we need to write one of the digits 1, 2, or 3. We need to do this in such a way that each of the digits 1, 2, and 3 will be written in each horizontal row and in each vertical column. If we start with 1 in the upper left cell, in how many different ways can the square be filled?

(A) 2   (B) 3   (C) 4   (D) 5   (E) 8

24. The weights in the figure are balanced. The same shapes have the same weight. The weight of each circular shape is 30 ounces. What is the weight in ounces of the square shape indicated by the question mark?

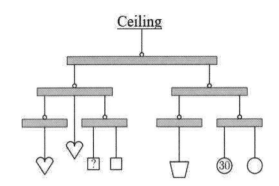

(A) 10   (B) 20   (C) 30   (D) 40   (E) 5

Problems from Year 2007

Problems 3 points each

1. Not taking any steps backwards, Anna travelled toward the car using a path shown in the picture, and picked up numbers she encountered along her way. Which set of the numbers below could she pick up?

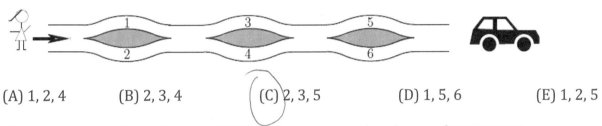

(A) 1, 2, 4   (B) 2, 3, 4   (C) 2, 3, 5   (D) 1, 5, 6   (E) 1, 2, 5

2. How many letters from the word KANGUR are repeated in the word PROBLEM?

(A) 1   (B) 2   (C) 3   (D) 4   (E) 5

3. Which of the patterns shown below consists of the largest number of squares?

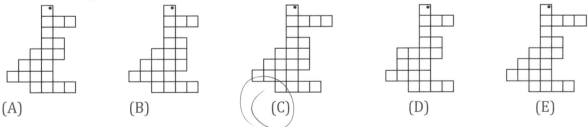

(A)   (B)   (C)   (D)   (E)

4. Helen has $5. She is going to buy 5 notebooks which cost 80 cents each and a certain number of pencils which cost 30 cents each. How many pencils at most can she buy?

(A) 5   (B) 4   (C) 3   (D) 2   (E) 1

5. There are 9 streetlights on one side of a path in the park. The distance between neighboring streetlights is 8 meters. Gregory went along this path from the first lantern to the last lantern. How many meters did he walk?

(A) 48   (B) 56   (C) 64   (D) 72   (E) 80

6. A 3-digit code is needed to open a safe. How many possible codes are there if it is known that only three numbers, 1, 3, and 5, are used in this code, and each of them is used only once?

(A) 2   (B) 3   (C) 4   (D) 5   (E) 6

7. 4 × 4 + 4 + 4 + 4 + 4 + 4 × 4 = ?

(A) 32   (B) 44   (C) 48   (D) 56   (E) 144

8. Which figure of those shown below can be placed next to the figure shown to the right in order form a rectangle?

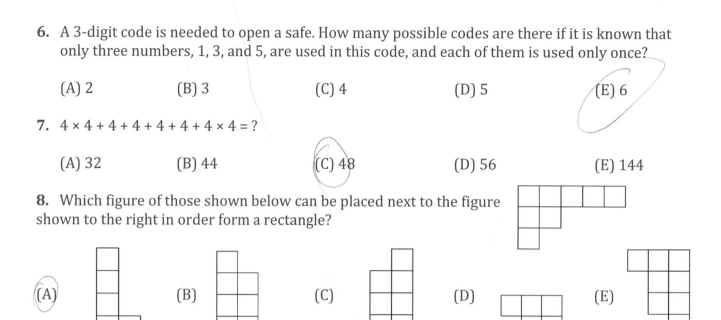

Problems 4 points each

9. What number needs to be written in the shaded cloud in order to get the number in the last cloud as the result of the operations shown in the picture?

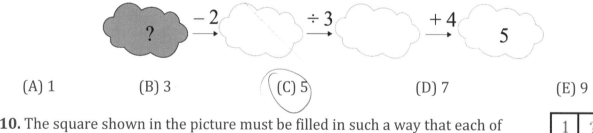

(A) 1   (B) 3   (C) 5   (D) 7   (E) 9

10. The square shown in the picture must be filled in such a way that each of the digits 1, 2, and 3 appears in each row and in each column once and only once. If Harry started to fill in the square as shown, what number can he write in the square marked with the question mark?

(A) 1   (B) 2   (C) 3   (D) 1 or 2   (E) 1, 2 or 3

11. What is the smallest number greater than 2007 in which the sum of the digits is equal to the sum of the digits of 2007?

(A) 2016   (B) 2015   (C) 2009   (D) 1008   (E) 2070

12. Annette is putting identical cube blocks into a cube aquarium. She has already put in a certain number of blocks (look at the picture). How many such blocks does she still need to add to fill up the aquarium?

    (A) 9   (B) 13   (C) 17   (D) 21   (E) 27

13. Peter, who is 1 year and 1 day older than Paul, was born on January 1st, 2002. When was Paul born?

    (A) January 2nd, 2003   (B) January 2nd, 2001   (C) December 31st, 2000
    (D) December 31st, 2002   (E) December 31st, 2003

14. A string has been cut into 400 pieces, each 15 cm long. How long was the string? (Note: 1 km = 1000 m, 1 m = 100 cm, 1 cm = 10 mm)

    (A) 6 km   (B) 60 m   (C) 600 cm   (D) 6000 mm   (E) 60,000 cm

15. David wrote a one-digit number, and next to it to the right he wrote another digit to form a two-digit number. Then he added 19 to this number and the sum was 72. What was the first digit he wrote?

    (A) 2   (B) 5   (C) 6   (D) 7   (E) 9

16. An electronic watch indicates the time 02:07. After how much time will the same digits show up again for the first time, not necessarily in the same order?

    (A) 4 hr 55 min   (B) 6 hr   (C) 10 hr 55 min   (D) 11 hr 13 min   (E) 24 hr

Problems 5 points each

17. A cube with an edge 3 cm long has been painted gray. Next, it has been cut into small cubes with an edge 1 cm long (see the picture). How many small cubes have exactly two gray sides?

    (A) 4   (B) 6   (C) 8   (D) 10   (E) 12

18. We call a number palindromic if it doesn't change after its digits are written in reverse order. Some examples are 1331 and 24642. A certain car's odometer shows 15951 kilometers. After how many more kilometers will a palindromic number show up on the odometer the very next time?

    (A) after 100 km   (B) after 110 km   (C) after 710 km   (D) after 900 km   (E) after 1010 km

19. If you count the small white squares in the sequence of big squares shown in the pictures below, you will get the numbers listed.

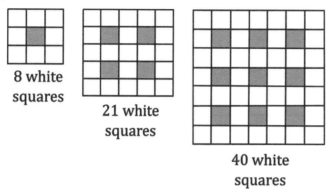

8 white squares

21 white squares

40 white squares

If we continue the pattern, how many white squares will there be in the next big square?

(A) 50   (B) 60   (C) 65   (D) 70   (E) 75

20. Adam, Bob, Celina, Daniel, and Eve formed a line to the cashier. Adam is standing farther from the cashier than Celina. Bob is standing closer to the cashier than Adam and right behind Daniel. Daniel is standing closer than Celina, but he isn't first in line. In what place, counting from the cashier, is Eve standing?

(A) 1st   (B) 2nd   (C) 3rd   (D) 4th   (E) 5th

21. In four corners of a rectangle with dimensions 15 cm × 9 cm, four squares each with a perimeter equal to 8 cm were cut out. What is the perimeter of the polygon created in this way?

(A) 48 cm   (B) 40 cm   (C) 32 cm   (D) 24 cm   (E) 16 cm

22. At a round table there are chairs placed with the same distance between them. They are numbered consecutively 1, 2, 3, … . Joe is sitting in the chair number 11, directly across from Chris, who is sitting in the chair number 4. How many chairs are there at the table?

(A) 13   (B) 14   (C) 16   (D) 17   (E) 22

23. How many digits have to be written in order to write down every number from 1 to 100 inclusive?

(A) 100   (B) 150   (C) 190   (D) 192   (E) 200

24. A square sheet of paper is folded twice so that a square is created again. In that square one of the corners is cut off (see the picture). Which of the pictures below cannot represent this sheet of paper once it is unfolded?

(A) (B) (C) (D)

(E) Each of the pictures (A), (B), (C), (D) can represent this unfolded sheet of paper.

Problems from Year 2008

Problems 3 points each

1. Ann eats 3 pieces of candy each day. How many pieces of candy does she eat in a week?

(A) 7   (B) 18   (C) 21   (D) 28   (E) 37

2. An adult ticket to the zoo costs $4, and a ticket for a child is $1 cheaper. On a certain Sunday, a father went to the zoo with his two children. How much did they have to pay for the tickets?

(A) $5   (B) $6   (C) $7   (D) $10   (E) $12

3. Luke gave the bouquets of flowers shown below to his mother, grandmother, aunt, and two sisters. Which of the bouquets did his mother receive, if we know that the flowers his aunt and sisters received were the same color, and that his grandmother did not receive roses?

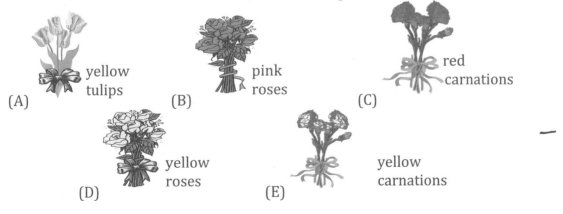

4. Adelaide has 37 CDs. Her friend Mary said, "If you give me 10 of your CDs, then we will have the same number of CDs." How many CDs did Mary have?

(A) 10   (B) 17   (C) 22   (D) 27   (E) 32

5. Jack drew a point on a piece of paper. Next, he drew four different straight lines going through this point. Into how many pieces did these lines divide the paper?

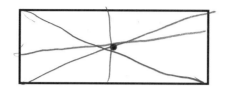

(A) 4   (B) 6   (C) 5   (D) 8   (E) 12

6. In 6 hours and 30 minutes the clock will show 4 o'clock. What time does the clock show now?

(A) 9:30   (B) 4:00   (C) 8:00   (D) 2:30   (E) 10:30

7. Charlie is playing with two identical cards which are equilateral triangles, as shown. He places them on a clean piece of paper, either partly on top of each other or touching each other, and then he traces the figure. Which figure can he not get in this way?

  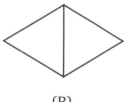 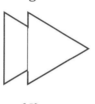  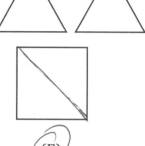

(A)   (B)   (C)   (D)   (E)

8. The storm made a hole in one side of the roof (see the picture). There were 10 roof tiles in each of the 7 rows. How many tiles are left on this side of the roof?

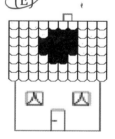

(A) 57   (B) 59   (C) 61   (D) 67   (E) 70

Problems 4 points each

9. In 2008, the Math Kangaroo competition is taking place in some school for the seventeenth time. Maggie took part in the seventh Math Kangaroo when she was 10 years old. In what year was Maggie born?

(A) 1986   (B) 1987   (C) 1998   (D) 1990   (E) 1988

10. Greg likes to multiply by 3, Jim likes to add 2, and Michael likes to subtract 1. In what order should the boys perform their favorite operations, each boy only once, so that starting with the number 3 they end up with 14?

(A) Greg, Jim, Michael   (B) Michael, Greg, Jim   (C) Greg, Michael, Jim
(D) Jim, Greg, Michael   (E) Michael, Jim, Greg

11. Grace is taller than Ann, but shorter than Tania. Irena is taller than Kate, but shorter than Grace. Which of the girls is the tallest?

   (A) Grace   (B) Ann   (C) Kate   (D) Irena   (E) Tania

12. Each of the figures A to E shown below is made out of 5 blocks. Which of the figures can you not get from the figure on the right if you move exactly one cube?

   (A)   (B)   (C)   (D)   (E)

13. For every two different numbers with the sum equal to 45, at least one of them is less than

   (A) 5.   (B) 18.   (C) 20.   (D) 22.   (E) 23.

14. A certain small hotel can house 21 guests. There are 5 three-person rooms and a certain number of two-person rooms. How many two-person rooms are there?

   (A) 1   (B) 2   (C) 3   (D) 4   (E) 6

15. There are three songs on a certain CD. The first song is 6 minutes and 25 seconds long, the second song is 12 minutes and 25 seconds long, and the third song is 10 minutes and 13 seconds long. How long does it take to play the whole CD?

   (A) 28 minutes 30 seconds   (B) 29 minutes 3 seconds   (C) 30 minutes 10 seconds
   (D) 31 minutes 13 seconds   (E) 31 minutes 23 seconds

16. Lynn shot 2 arrows at the target and got 5 points—see the picture. How many different scores can one get if both arrows hit the target?

   (A) 4   (B) 6   (C) 8   (D) 9   (E) 10

Problems 5 points each

17. Blocks with all right angles with the dimensions 1 cm × 2 cm × 4 cm are packed into cubes with the dimensions 4 cm × 4 cm × 4 cm. How many blocks make one of these cubes?

   (A) 6   (B) 7   (C) 8   (D) 9   (E) 10

18. A kangaroo noticed that each winter he gains 5 kilograms of weight, and each summer he loses 4 kilograms. During the spring and fall, his weight does not change. In the spring of 2008, he weighed 100 kilograms. How much did he weigh in the fall of 2004?

(A) 92 kilograms   (B) 93 kilograms   (C) 94 kilograms   (D) 96 kilograms   (E) 98 kilograms

19. A square garden was divided into four parts: pool (P), flowerbed (F), lawn (L), and sandbox (S)—see the picture. The lawn and flowerbed are squares. The perimeter of the lawn is 20 m, and the perimeter of the flowerbed is 12 m. What is the perimeter of the pool?

(A) 10 m   (B) 12 m   (C) 14 m   (D) 16 m   (E) 18 m

20. A boy named Henry has as many brothers as sisters. His sister Diane has twice as many brothers as sisters. How many children are there in the family?

(A) 3   (B) 4   (C) 5   (D) 6   (E) 7

21. How many two-digit numbers are there where the ones digit is greater than the tens digit?

(A) 26   (B) 18   (C) 9   (D) 30   (E) 36

22. Right now, Mary is five times as old as her sister Li. In 6 years, she will be twice as old as Li. How old will Mary be in 10 years?

(A) 15   (B) 20   (C) 25   (D) 30   (E) 35

23. In the botanical garden shown in the picture, visitors walk only on the marked paths. In how many different ways can one go from greenhouse A to greenhouse B if you only walk on a given path once?

(A) 4   (B) 6   (C) 8   (D) 10   (E) 12

24. Altogether, there are 40 liters of water in two containers. First, 5 liters were poured from the first container to the second, and then enough water was poured from the second container to the first to double the amount of water in the first container. It turned out that at that point both containers ended up with the same amount of water. How much water was in the second container at the beginning?

(A) 20   (B) 35   (C) 15   (D) 25   (E) 10

Problems from Year 2009

Problems 3 points each

1. The figure shown in the picture was made out of identical wooden cubes. How many wooden cubes were used?

    (A) 6   (B) 8   (C) 10   (D) 12   (E) 15

    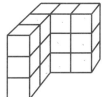

2. 200 × 9 + 200 + 9 =

    (A) 418   (B) 1909   (C) 2009   (D) 4018   (E) 20009

3. Where is the kangaroo?

    (A) In the circle and in the triangle, but not in the square.
    (B) In the circle and in the square, but not in the triangle.
    (C) In the triangle and in the square, but not in the circle.
    (D) In the circle, but not in the square and not in the triangle.
    (E) In the square, but not in the circle and not in the triangle.

    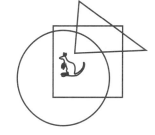

4. In a certain family there are five brothers. Each one of them has one sister. How many children are in this family?

    (A) 6   (B) 7   (C) 8   (D) 9   (E) 10

5. The number 930 is shown on a display (see the picture). How many little squares need to change color in order to show the number 806?

    (A) 5   (B) 6   (C) 7   (D) 8   (E) 9

6. Mother bought 16 oranges. Carl ate half of them, Eva ate two, and Sophie ate the rest. How many oranges did Sophie eat?

    (A) 4   (B) 6   (C) 8   (D) 10   (E) 12

7. In his garden Anthony made the path shown in the figure, using 10 tiles of size 4 m by 6 m. Anthony painted a black line between the midpoints of the tiles. How long is this black line?

    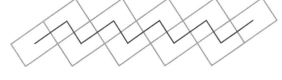

    (A) 24 m   (B) 40 m   (C) 46 m   (D) 50 m   (E) 56 m

© Math Kangaroo in USA, NFP   www.mathkangaroo.org

8. A certain movie is 90 minutes long. It started at 5:10 p.m. During the movie, there were two commercial breaks, one lasting 8 minutes and one lasting 5 minutes. At what time did the movie finish?

   (A) at 6:13 p.m.   (B) at 6:27 p.m.   (C) at 6:47 p.m.   (D) at 6:53 p.m.   (E) at 7:13 p.m.

Problems 4 points each

9. A red kangaroo and a gray kangaroo together weigh 139 kilograms. The red kangaroo weighs 35 kilograms less than then gray kangaroo. How much does the gray kangaroo weigh?

   (A) 104 kilograms   (B) 52 kilograms   (C) 87 kilograms
   (D) 96 kilograms   (E) 53 kilograms

10. Zach was dividing a chocolate bar. He broke one row of five pieces for his brother and then one row of seven pieces for his sister, as shown in the picture. How many pieces were there in the whole chocolate bar?

    (A) 28   (B) 32   (C) 35   (D) 40   (E) 54

11. A certain dance group started out with 25 boys and 19 girls. Every week 2 more boys and 3 more girls join the dance group. After how many weeks will there be the same number of boys and girls in the dance group?

    (A) 6   (B) 5   (C) 4   (D) 3   (E) 2

12. A farmer has 30 cows, some chickens, and no other animals. The total number of the legs of the chickens is equal to the total number of the legs of the cows. How many animals does the farmer have altogether?

    (A) 60   (B) 90   (C) 120   (D) 180   (E) 240

13. One side of a rectangle is 8 cm long, while the other is half as long. A square has the same perimeter as the rectangle. What is the length of the side of the square?

    (A) 4 cm   (B) 6 cm   (C) 8 cm   (D) 12 cm   (E) 24 cm

14. Magda rolled a die four times and she obtained a total of 23 points. How many times did the roll show 6 dots?

    (A) 0   (B) 1   (C) 2   (D) 3   (E) 4

15. Three squirrels, Hela, Mela and Tola, together found 7 nuts. Each of them found a different number of nuts, but each of them found at least one nut. Hela collected the least, and Mela the most of all. How many nuts did Tola find?

(A) 1   (B) 2   (C) 3   (D) 4   (E) 5

16. Peter and Paul went to a boy scout camp. During a meeting, the scouts stood in a single row. On one side of Paul there were 27 scouts, and on the other side there were 13 scouts. Peter was standing exactly in the middle of the row. How many scouts were there between Peter and Paul?

(A) 6   (B) 7   (C) 8   (D) 14   (E) 21

Problems 5 points each

17. Which of the figures below cannot be made using the two dominoes shown in the picture to the right?

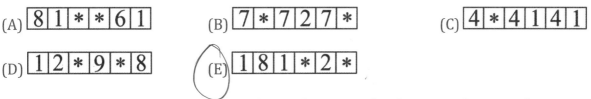

(A)   (B)   (C)   (D)   (E)

18. A secret agent wants to break a 6-digit code. He knows that the sum of the first, third, and fifth digits is equal to the sum of the second, fourth, and sixth digits. Which of the following could be the code?

(A) 81**61   (B) 7*727*   (C) 4*4141

(D) 12*9*8   (E) 181*2*

19. One week, Ms. Florentina sold eggs at the market every day from Monday to Friday. On Wednesday, she sold 60 eggs. On Thursday, she sold 96 eggs, and noticed that every day that week the number of eggs she sold was equal to the sum of the number of eggs she sold the two previous days. How many eggs did Ms. Florentina sell on Monday?

(A) 20   (B) 24   (C) 36   (D) 40   (E) 48

20. A certain vase contains four flowers: one red, one blue, one yellow, and one white. Kaya the Bee sat on every flower in the bouquet only once. She started with the red flower, and she did not fly directly from the yellow flower to the white flower. In how many ways could Maia sit on all the flowers?

(A) 1 (B) 2 (C) 3 (D) 4 (E) 6

21. At 6:15 a.m. Jasper the Friendly Ghost vanished, and the crazy clock, which had been showing the right time until then, started to run at the right speed but backwards. The ghost appeared again at 7:30 p.m. that same day. What time did the crazy clock show at the moment when the ghost reappeared?

(A) 5:00 p.m. (B) 5:45 p.m. (C) 6:30 p.m. (D) 7:00 p.m. (E) 7:15 p.m.

22. The squares of a 3 × 3 table were filled in with numbers as shown in the picture. In one move, we can switch any two numbers. What is the smallest number of such moves that we need to make to get a table in which the sum of the numbers in each row is divisible by 3?

4	5	1
8	10	4
7	1	2

(A) 1 (B) 2 (C) 3 (D) 4 (E) It is impossible to get such a table.

23. Agnes was drawing figures made out of segments of length 1. At the end of each segment, she always turned at a right angle either to the left or to the right. Each time she turned right, she drew the symbol ♥ on a piece of paper, and each time she turned left, she drew the symbol ♠. One day, she drew a figure and drew these symbols in this order: ♥ ♠ ♠ ♠ ♥ ♥. Which of the following figures could Agnes have drawn?

(A) (B) (C) (D) (E)

24. In the land of Funnyfeet, the left foot of each man is two sizes bigger than his right foot, and the left foot of each woman is one size bigger than her right foot. However, shoes are always sold in pairs of the same size, and only in whole sizes. A group of friends decided to buy green shoes, and to save money they bought shoes together. After they all put on the shoes that fit them, there were exactly two shoes left over, one of size 36 and another of size 45. What is the smallest possible number of people in the group?

(A) 4 (B) 5 (C) 6 (D) 7 (E) 9

Problems from Year 2010

Problems 3 points each

1. Which of the numbers below is the greatest?

   (A) 2 + 0 − 1 + 0   (B) 2 − 0 − 1 + 0   (C) 2 + 0 − 1 − 0   (D) 2 − 0 + 1 + 0   (E) 2 − 0 − 1 − 0

2. A dance lesson lasts 40 minutes, and it started at 11:50. Stan was late and came to class exactly half way through. What time did Stan come to class?

   (A) 11:30   (B) 12:00   (C) 12:10   (D) 12:20   (E) 12:30

3. Several kids were measuring the length of a sandbox by their steps. Anna needed 15 whole steps to walk its length, Beata needed 17 whole steps, Daniel needed 12 whole steps, and Igor needed 14 whole steps. Whose steps were the longest?

   (A) Anna's   (B) Beata's   (C) Daniel's   (D) Igor's   (E) It cannot be determined.

4. Adam spent five days preparing for a test. The first day he solved one problem, and on each consecutive day he solved twice as many problems as the day before. How many problems did Adam solve altogether preparing for the test?

   (A) 15   (B) 16   (C) 31   (D) 33   (E) 63

5. John has 4 pieces of cardboard with this design: . Which of the patterns below can he not make using these pieces?

   (A)   (B)    (C) 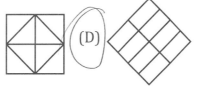   (D)   (E)

6. At the cosmetics store, you can buy a bar of soap for $4, a bottle of shampoo for $9, and jar of face cream for $5. You can also buy these three items as a set for $15. How much money would mother save if she bought the set instead of buying the three items separately?

   (A) $3   (B) $4   (C) $5   (D) $6   (E) $7

7. Eva the centipede has 50 pairs of feet. She had shoes on some of her feet, but on the rest of her feet she did not have shoes. Today she bought 16 pairs of new shoes and put them on the feet that were without shoes. She still has 7 pairs of feet without shoes. On how many feet did she have shoes before she bought the 16 pairs of shoes?

   (A) 27   (B) 40   (C) 54   (D) 70   (E) 77

PROBLEMS 2010

8. The figure shown in Picture 1 was made out of six identical coins. What is the smallest number of coins that we need to move to make the figure shown in Picture 2?

(A) 1   (B) 2   (C) 3   (D) 4   (E) 5

Picture 1

Picture 2

Problems 4 points each

9. The product 60 × 60 × 24 × 7 is equal to

(A) the number of minutes in seven weeks.
(B) the number of seconds in seven hours.
(C) the number of minutes in twenty-four weeks.
(D) the number of hours in sixty days.
(E) the number of seconds in one week.

10. Adam, Luke, Tom and Alex went out for ice cream. Adam ate more ice cream than Alex, and Tom ate more ice cream than Luke but less than Alex. Which of the lists below gives the names of the boys in order from the one who ate the most ice cream to the one who ate the least?

(A) Adam, Tom, Luke, Alex   (B) Adam, Alex, Tom, Luke
(C) Tom, Adam, Luke, Alex   (D) Alex, Adam, Luke, Tom
(E) Tom, Luke, Adam, Alex

11. Matthew and Clara live in a skyscraper. Clara lives 12 floors above Matthew. One day Matthew went to visit Clara, and he took the stairs up from his apartment to Clara's apartment. Half-way up he was on the 8th floor. On what floor does Clara live?

(A) 12   (B) 14   (C) 16   (D) 20   (E) 24

12. We made a big cube using 64 small white cubes, and then we painted five of the sides of the big cube. How many of the small cubes have exactly two sides painted?

(A) 4   (B) 8   (C) 16   (D) 20   (E) 24

13. Grandpa has a 10-meter by 20-meter yard. One of the shorter sides of the yard is along the house. Grandpa decided to put a fence around the remaining three sides of the yard and to paint the fence. Painting 5 meters of the fence uses one-half of a two-gallon can of paint. A can of paint costs $40. How much will Grandpa pay for the paint he needs to paint the whole fence?

(A) $120   (B) $400   (C) $240   (D) $200   (E) $480

14. Camilla wrote all the positive integers from 1 to 100 in order on a chart with 5 columns. A part of the chart is shown in the picture to the right. Her brother cut out a part of the chart and then he erased some of the numbers from it. Which picture represents the part of the incomplete chart cut out by Camilla's brother?

1	2	3	4	5
6	7	8	9	10
11	12	13	14	15
16	17	18	19	20

(A) | | 43 | | |
    | | | 48 | |

(B) | | | | 60 |
    | | 52 | | |

(C) | | | 69 | |
    | 72 | | | |

(D) | | 81 | | |
    | 86 | | | |

(E) | | 87 | | |
    | | | 94 | |

15. A square piece of paper is white on one side and green on the other side. Anne divided it into 9 little squares. She labeled some edges with natural numbers 1 to 8 (see Picture 1). What is the sum of the numbers along the edges which she cut (see Picture 2)?

(A) 16 (B) 17 (C) 18 (D) 20 (E) 21

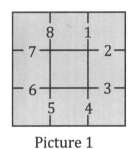
Picture 1

Picture 2

16. If the numbers in each of the two rows have the same sum, what is the value of *?

1	2	3	4	5	6	7	8	9	10	199
11	12	13	14	15	16	17	18	19	20	*

(A) 99   (B) 100   (C) 209   (D) 289   (E) 299

Problems 5 points each

17. Ella folded a square piece of paper twice, and so made a square with each of its sides as long as half of the original piece of paper. She then cut off all four corners from the square she made. Which of the pictures below shows the piece of paper after unfolding?

(A)   (B)   (C)   (D) 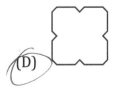   (E)

18. Anna, Beata, and Jack go to the same school. One day the librarian said to them, "Guess how many books we have in the school library." Anna said 2010, Beata said 1998, and Jack said 2015. It turned out that the number of books in the library differed from the numbers given by the children by 12, 7 and 5 (these numbers are not necessarily in the order they made their guesses). How many books are there in their school library?

(A) 2005   (B) 2008   (C) 2003   (D) 2020   (E) 2022

19. Adam and Tom are walking in the same direction around a circular table and counting chairs. They begin their count with different chairs. Tom's twelfth chair is Adam's third chair, while Tom's fifth chair is Adam's eighteenth chair. How many chairs are there at the table?

(A) 20 (B) 18 (C) 30 (D) 22 (E) 23

20. A jeweler can make chains of any length using identical links. Picture 1 shows such a chain made out of three links. A single link is shown in Picture 2. What is the length of a chain made out of five links?

(A) 20 mm (B) 19 mm (C) 17.5 mm
(D) 16 mm (E) 15 mm

21. The ladybugs which live on an enchanted meadow are either red and have 6 spots, or yellow and have 10 spots. Various animals, including both red and yellow ladybugs, came to a birthday party for the dragonfly. The dragonfly noticed that the total number of spots on the ladybugs that were at the party was 42. How many ladybugs came to the dragonfly's birthday party?

(A) 10 (B) 7 (C) 6 (D) 8 (E) 5

22. Each of Basil's friends added the number of the day and the number of the month of his birthday and obtained 35. Their birthdays all fall on different days. What is the greatest possible number of friends that Basil has?

(A) 7 (B) 8 (C) 9 (D) 10 (E) 12

23. Paul, Darius, Micah, and Jeff met at a concert in Chicago, but they came from different cities: Pittsburgh, Dallas, New York, and Washington. We know that:
- Paul and the boy from Washington met in Chicago early in the morning on the day of the concert. They had never been to Pittsburgh or to New York.
- Micah is not from Washington, and he came to Chicago later than the boy from Pittsburgh.
- Jeff liked the concert more than the boy from Pittsburgh did.

What city is Jeff from?

(A) Pittsburgh (B) New York (C) Dallas (D) Washington (E) Chicago

24. Olivia is ten years old and is six times younger than her grandmother. Olivia's grandmother is 14 years older than the ages of Olivia and of her mother added together. Olivia's great-grandmother's age is equal to the sum of Olivia's grandmother's and Olivia's mother's ages. How old is Olivia's great-grandmother?

(A) 106 (B) 69 (C) 70 (D) 89 (E) 96

Problems from Year 2011

Problems 3 points each

1. Which of the numbers below is the greatest?

   (A) 20 + 11   (B) 20 − 11   (C) 20 + 1 + 1   (D) 20 − 1 − 1   (E) 2 + 0 + 1 + 1

2. Michael is painting the word KANGAROO on a poster. Each day he paints one letter. He painted the first letter on a Wednesday. What day of the week will it be when he paints the last letter?

   (A) Monday   (B) Tuesday   (C) Wednesday   (D) Thursday   (E) Friday

3. Which of the stones below needs to be added to the box on the right side of the scale in order for the two boxes on the scale to weigh the same? (The numbers on the stones show their weights in kilograms.)

   (A) 5   (B) 7   (C) 9
   (D) 11   (E) 13

   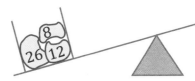

4. The train to Atlanta leaves three and a half hours from now. Paul got up two and a half hours ago. How many hours before the train leaves did Paul get up?

   (A) two and a half   (B) three and a half   (C) four and a half   (D) five   (E) six

5. A toy is placed in one of the squares of a grid, as shown in the picture. A child moved the toy from one square to the next. He first moved it one square to the right, then one square up, then one square to the left, then one square down, and then again one square to the right. Which of the following pictures shows where the toy was in the end?

   (A)    (B) 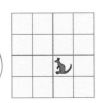   (C)   (D)    (E)

6. Ala, Lenka, and Miso went out for dessert. Lenka paid 4 dollars and 50 cents for three scoops of ice cream. Miso paid 3 dollars and 60 cents for two cookies. How much did Ala pay for one scoop of ice cream and one cookie?

   (A) 3 dollars and 30 cents   (B) 4 dollars and 80 cents   (C) 5 dollars and 10 cents
   (D) 6 dollars and 30 cents   (E) 8 dollars and 10 cents

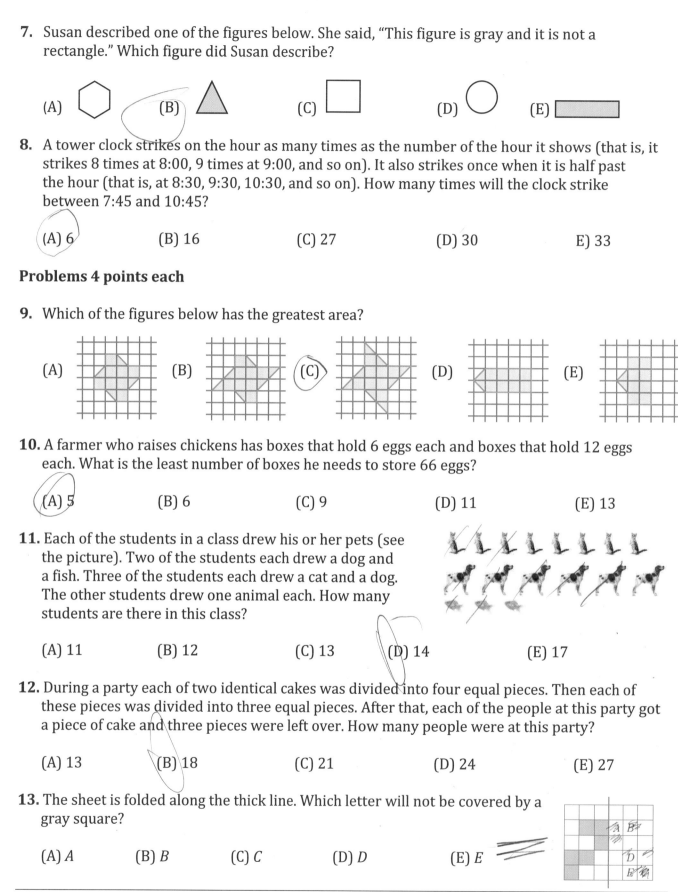

14. There are 13 coins in John's pocket, and each of them is either a 5-cent coin or a 10-cent coin. Which of the numbers below cannot be the total value of John's coins?

(A) 60 cents   (B) 70 cents   (C) 80 cents   (D) 115 cents   (E) 125 cents

15. Ari, Chuck, Darius, Jack, Mark, and Tom were rolling a six-sided die. Each of the boys rolled the die once and each one got a different number. Ari got a result four times as great as Chuck. Darius got a result twice as great as Jack and three times as great as Mark. What number did Tom roll?

(A) 2   (B) 3   (C) 4   (D) 5   (E) 6

16. A squirrel went through the maze gathering nuts (see the picture). It could go only once through each door between the rooms of the maze. What is the greatest number of nuts it could have gathered?

(A) 7   (B) 10   (C) 11   (D) 12   (E) 15

Problems 5 points each

17. A certain quiz show has the following rules: Every participant has 10 points at the beginning and has to answer 10 questions. For each correct answer, the participant earns 1 point, and for each incorrect answer, the participant loses 1 point. Mrs. Smith had 14 points at the end of this quiz show. How many correct answers did she give?

(A) 7   (B) 8   (C) 9   (D) 6   (E) 4

18. Four friends, Masha, Sasha, Dasha, and Pasha, were sitting on a bench. First Masha changed places with Dasha. Then Dasha changed places with Pasha. At the end the girls sat on the bench in the following order from left to right: Masha, Sasha, Dasha, Pasha. In what order from left to right were they sitting in the beginning?

(A) Masha, Sasha, Dasha, Pasha   (B) Masha, Dasha, Pasha, Sasha
(C) Dasha, Sasha, Pasha, Masha   (D) Sasha, Masha, Dasha, Pasha
(E) Pasha, Masha, Sasha, Dasha

19. The diagram shows the distances in miles between certain towns A, B, C, D, E, and F. What is the distance in miles between towns C and D?

(A) 5   (B) 6   (C) 10   (D) 11   (E) 13

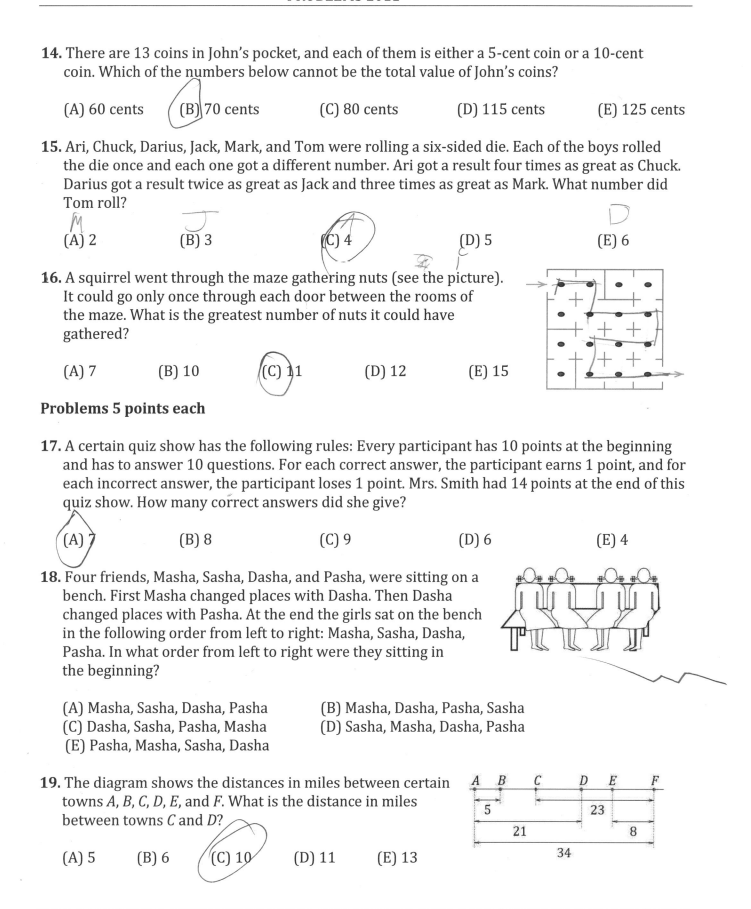

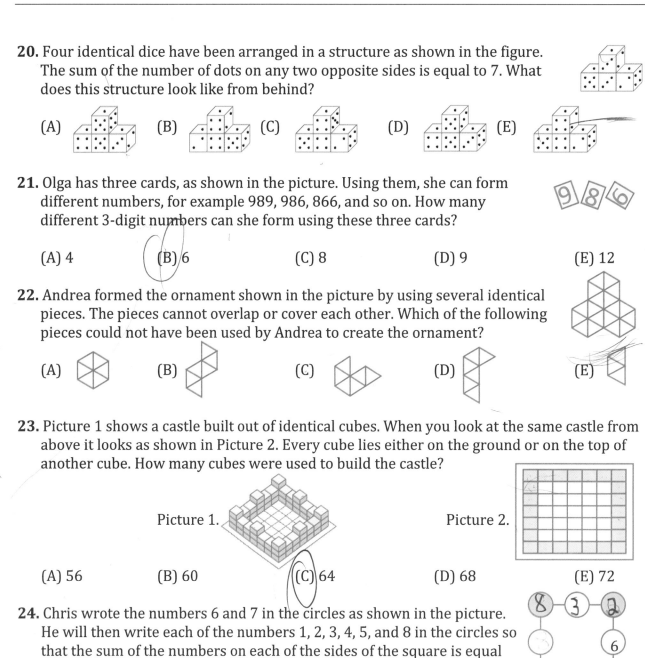

Problems from Year 2012

Problems 3 points each

1. Basil wants to write the word MATHEMATICS on a sheet of paper. He wants to color different letters with different colors, and the same letters with the same color. How many colors will he need?

   (A) 7 (B) 8 (C) 9 (D) 10 (E) 13

2. In four of the five pictures below the white area is equal to the gray area. In which picture are the white area and the gray area different?

   (A)   (B)   (C)   (D)   (E)

3. Father hangs the laundry outside on a clothesline. He wants to use as few pins as possible. For 3 towels he needs 4 pins, as shown. How many pins does he need for 9 towels?

   (A) 8 (B) 10 (C) 12 (D) 14 (E) 16

   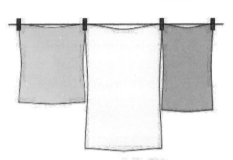

4. Iljo colors the squares A2, B1, B2, B3, B4, C3, D3 and D4 gray. Which pattern does he get?

   (A) 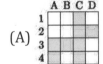 (B) 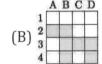 (C) 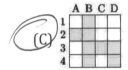 (D) 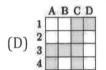 (E)

5. 13 children are playing hide and seek. One of them is the "seeker" and the others hide. After a while, 9 children have been found. How many children are still hiding?

   (A) 3 (B) 4 (C) 5 (D) 9 (E) 22

6. Mike and Jake were playing darts. Each one threw three darts (see the picture). Who won and how many points more than his opponent did he score?

    (A) Mike; he scored 3 points more.
    (B) Jake; he scored 4 points more.
    (C) Mike; he scored 2 points more.
    (D) Jake; he scored 2 points more.
    (E) Mike; he scored 4 points more.

7. A regular rectangular pattern on a wall was created with 2 kinds of tiles: gray and striped. Some tiles have fallen off the wall (see the picture). How many gray tiles have fallen off?

    (A) 9   (B) 8   (C) 7   (D) 6   (E) 5

8. The year 2012 is a leap year, which means that there are 29 days in February. Today, on March 15, 2012, my grandfather's ducklings are 20 days old. When did they hatch from their eggs?

    (A) on February 19   (B) on February 21   (C) on February 23
    (D) on February 24   (E) on February 26

Problems 4 points each

9. You have L-shaped tiles, each consisting of 4 squares as shown: How many of the following shapes can you make by gluing together two of these tiles?

    (A) 0   (B) 1   (C) 2   (D) 3   (E) 4

10. Three balloons cost 12 cents more than one balloon. How many cents does one balloon cost?

    (A) 4   (B) 6   (C) 8   (D) 10   (E) 12

11. Grandmother made 20 gingerbread cookies for her grandchildren. She decorated them with raisins and nuts. First, she decorated 15 cookies with raisins and then she decorated 15 cookies with nuts. At least how many cookies were decorated with both raisins and nuts?

    (A) 4   (B) 5   (C) 6   (D) 8   (E) 10

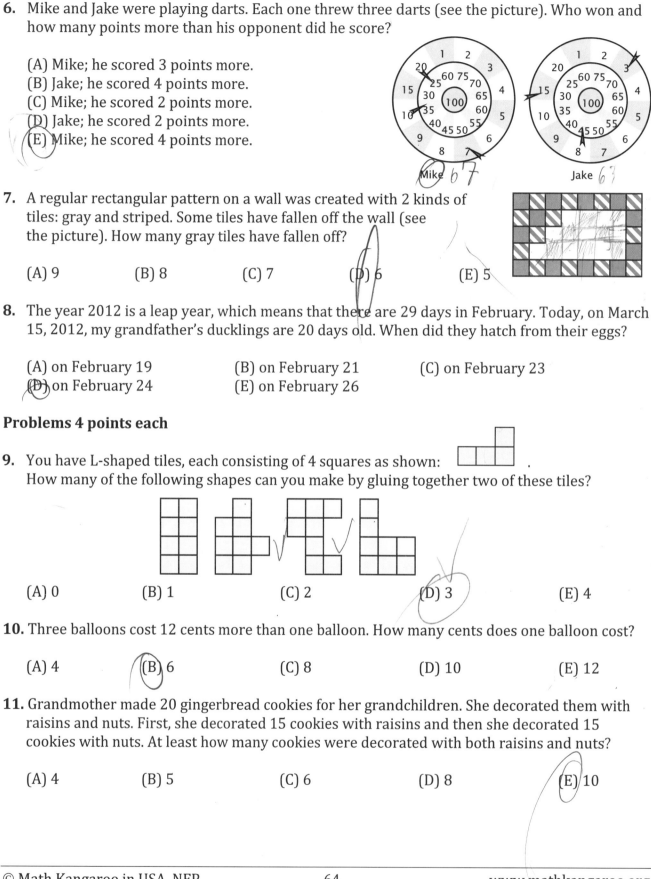

PROBLEMS 2012

12. In a sudoku puzzle the numbers 1, 2, 3, 4 can occur only once in each column and in each row. In the mathematical sudoku to the left, Patrick first writes in the results of the calculations. Then he completes the sudoku. Which number will Patrick put in the gray cell?

1×1		1×3	
2×2	$6 - 3$		$6 - 5$
$4 - 1$	$1 + 3$	$8 - 7$	
$9 - 7$	$2 - 1$		

(A) 1   (B) 2   (C) 3   (D) 4   (E) 1 or 2

13. Among Nikolay's classmates there are twice as many girls as boys. Which of the following numbers can be equal to the number of all the children in this class?

(A) 30   (B) 20   (C) 24   (D) 25   (E) 29

14. In the animal school, 3 kittens, 4 ducklings, 2 baby geese, and several lambs are taking lessons. The teacher owl found out that all of her pupils have 44 legs altogether. How many lambs are there among them?

(A) 6   (B) 5   (C) 4   (D) 3   (E) 2

15. A rectangular prism is made of four pieces, as shown. Each piece consists of four cubes and is a single color. What is the shape of the white piece?

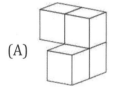 (A)   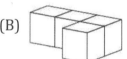 (B)   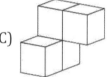 (C)   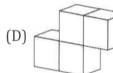 (D)   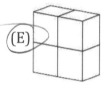 (E)

16. At a Christmas party there was exactly one candlestick on each of the 15 tables. There were 6 five-branched candlesticks; the rest of them were three-branched candlesticks. How many candles had to be bought for all the candlesticks?

(A) 45   (B) 50   (C) 57   (D) 60   (E) 75

Problems 5 points each

17. A grasshopper wants to climb a staircase with many steps. She makes only two kinds of jumps: 3 steps up or 4 steps down. Beginning at the ground level, at least how many jumps will she have to make in order to take a rest on the 22th step?

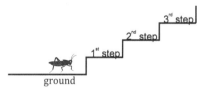

(A) 7   (B) 9   (C) 10   (D) 12   (E) 15

© Math Kangaroo in USA, NFP   www.mathkangaroo.org

18. Frank made a domino snake out of seven tiles. He put the tiles next to each other so that the sides with the same number of dots were touching. Originally the snake had 33 dots on its back. However, his brother George took away two tiles from the snake (see the picture). How many dots were in the place with the question mark?

(A) 2   (B) 3   (C) 4   (D) 5   (E) 6

19. Gregory forms two numbers with the digits 1, 2, 3, 4, 5, and 6. Both numbers have three digits, and each digit is used only once. He adds these two numbers. What is the greatest sum Gregory can get?

(A) 975   (B) 999   (C) 1083   (D) 1173   (E) 1221

20. Laura, Iggy, Val, and Kate want to be in one photo together. Kate and Laura are best friends and they want to stand next to each other. Iggy wants to stand next to Laura because he likes her. In how many different ways can they pose for the photo if they all stand in one row?

(A) 3   (B) 4   (C) 5   (D) 6   (E) 7

21. A special clock has 3 hands of different length (one for hours, one for minutes, and one for seconds). We do not know which hand is which, but we know that the clock runs correctly. At 12:55:30 p.m. the hands were in position shown on the right. What will this clock look like at 8:11:00 p.m.?

(A)   (B)   (C)   (D)   (E)

22. Michael chose a positive number, multiplied it by itself, added 1, multiplied the result by 10, added 3, and multiplied the result by 4. His final answer was 2012. What number did Michael choose?

(A) 11   (B) 9   (C) 8   (D) 7   (E) 5

23. A rectangular paper sheet measures 192 × 84 mm. You cut the sheet along just one straight line to get two parts, one of which is a square. Then you do the same with the non-square part of the sheet, and so on. What is the length of the side of the smallest square you can get in this way?

(A) 1 mm   (B) 4 mm   (C) 6 mm   (D) 10 mm   (E) 12 mm

24. In a soccer game the winner gains 3 points, while the loser gains 0 points. If the game is a tie, then the two teams gain 1 point each. A certain team played 38 games and gained 80 points. Find the greatest possible number of games that the team lost.

   (A) 12   (B) 11   (C) 10   (D) 9   (E) 8

Problems from Year 2013

Problems 3 points each

1. In which figure is the number of black kangaroos larger than the number of white kangaroos?

   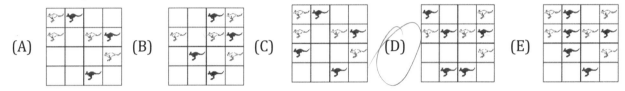

2. Aline writes a correct calculation. Then she covers two digits that are the same with stickers (see the picture). Which digit is under the stickers?

   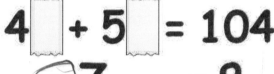

   (A) 2   (B) 4   (C) 5   (D) 7   (E) 8

3. In what way should the last four circles be shaded so that the pattern is continued?

   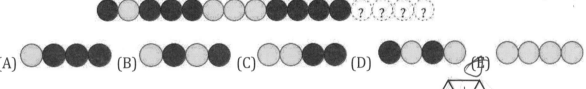

4. How many triangles can be seen in the picture to the right?

   (A) 9   (B) 10   (C) 11   (D) 13   (E) 12

5. In the London 2012 Olympics, USA won the most medals: 46 gold, 29 silver, and 29 bronze. China won the second most medals with 38 gold, 27 silver, and 23 bronze medals. How many more medals did USA win than China?

   (A) 6   (B) 14   (C) 16   (D) 24   (E) 26

6. Daniel had a package of 36 pieces of candy. Without breaking any pieces of candy, he divided all the candy equally among his friends. Which of the following was definitely not the number of his friends?

   (A) 2   (B) 3   (C) 4   (D) 5   (E) 6

7. Vero's mom prepares sandwiches with two slices of bread each. A package of bread has 24 slices. How many sandwiches can she prepare from two and a half packages of bread?

   (A) 24   (B) 30   (C) 48   (D) 34   (E) 26

8. About the number 325, five boys said:
     Andy: "This is a 3-digit number."
     Barry: "All the digits are different."
     Charlie: "The sum of the digits is 10."
     Danny: "The ones digit is 5."
     Eddie: "All the digits are odd."
   Which of the boys was wrong?

   (A) Andy   (B) Barry   (C) Charlie   (D) Danny   (E) Eddie

Problems 4 points each

9. A rectangular mirror was broken. Which of the following pieces is missing from the picture of the broken mirror?

   (A)   (B)   (C)   (D)   (E)

10. Each time Pinocchio lies, his nose gets 6 cm longer. Each time he tells the truth, his nose gets 2 cm shorter. After his nose was 9 cm long, he told three lies and made two true statements. How long was Pinocchio's nose afterwards?

    (A) 14 cm   (B) 15 cm   (C) 19 cm   (D) 23 cm   (E) 31 cm

11. In a shop you can buy oranges in boxes of three different sizes: boxes of 5 oranges, boxes of 9 oranges, or boxes of 10 oranges. Pedro wants to buy exactly 48 oranges. What is the smallest number of boxes he can buy?

    (A) 8   (B) 7   (C) 6   (D) 5   (E) 4

12. Ann starts walking in the direction of the arrow. At every intersection of the streets she turns either to the right or to the left. First she goes to the right, then to the left, then again to the left, then to the right, then to the left, and finally again to the left. Then Ann is finally walking towards

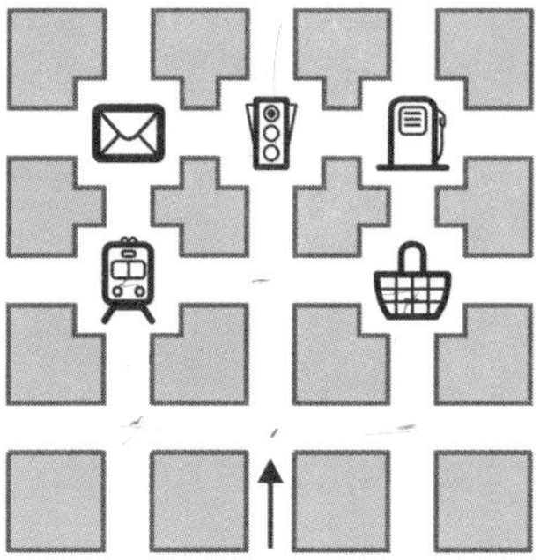

(A)  🚆 (B)  ✉ (C)  🚦 (D)  ⛽ (E)  🧺

13. Classmates Andy, Betty, Cathie, and Dannie were born in the same year. Their birthdays were on February 20th, April 12th, May 12th, and May 25th, not necessarily in this order. Betty and Andy were born in the same month. Andy and Cathie were born on the same day of different months. Who of these classmates is the oldest?

(A) Andy (B) Betty (C) Cathie (D) Dannie (E) It is impossible to determine.

14. At Adventure Park, each of 30 children took part in at least one of two events. 15 of them took part in the "moving bridge" contest, and 20 of them went down the zipline. How many of the children took part in both events?

(A) 25 (B) 15 (C) 30 (D) 10 (E) 5

15. Which of the following pieces fits with the piece in the picture to the right so that together they form a rectangle?

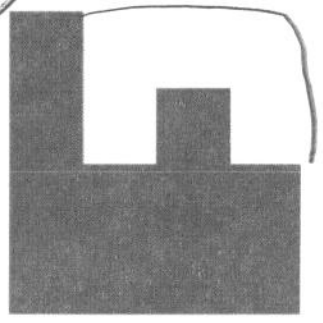

(A)  (B)  (C)  (D) (E)

16. The number 35 has the property that it is divisible by the digit in the ones position, because 35 divided by 5 is 7. The number 38 does not have this property. How many numbers greater than 21 and smaller than 30 have this property?

(A) 2 (B) 3 (C) 4 (D) 5 (E) 6

Problems 5 points each

17. Joining the midpoints of the sides of the triangle in the drawing, we obtain a smaller triangle. We repeat this one more time with the smaller triangle. How many triangles of the same size as the smallest resulting triangle fit in the original drawing?

(A) 5 (B) 8 (C) 10 (D) 16 (E) 32

18. After the first of January 2013, how many years will pass before the following event happens for the first time: the product of the digits in the notation of the year is greater than the sum of these digits?

(A) 87 (B) 98 (C) 101 (D) 102 (E) 103

19. In December Tosha-the-Cat slept for exactly 3 weeks. How many minutes did she stay awake during this month?

(A) $(31 - 7) \times 3 \times 24 \times 60$ (B) $(31 - 7 \times 3) \times 24 \times 60$ (C) $(30 - 7 \times 3) \times 24 \times 60$
(D) $(31 - 7) \times 24 \times 60$ (E) $(31 - 7 \times 3) \times 24 \times 60 \times 60$

20. Basil has several domino tiles, as shown below. He wants to arrange them in a line according to the following "domino rule": in any two neighboring tiles, the neighboring squares must have the same number of dots. What is the largest number of tiles he can arrange in this way?

(A) 3 (B) 4 (C) 5 (D) 6 (E) 7

21. Cristi has to sell 10 glass bells which vary in price: 1 dollar, 2 dollars, 3 dollars, 4 dollars, 5 dollars, 6 dollars, 7 dollars, 8 dollars, 9 dollars, and 10 dollars. In how many ways can Cristi divide all the glass bells into three packages so that each of the packages has the same price?

(A) 1 (B) 2 (C) 3 (D) 4 (E) Such division is not possible.

22. Peter bought a rug 36 in wide and 60 in long. The rug has a pattern of small squares containing either a sun or a moon, as can be seen in the figure. You can see that along the width there are 9 squares. When the rug is fully unrolled, how many moons can be seen?

(A) 68 (B) 67 (C) 65 (D) 63 (E) 60

23. Baby Roo wrote down as few numbers as possible using only the digits 0 and 1 to get 2013 as the sum. How many numbers did Baby Roo write?

(A) 2 (B) 3 (C) 4 (D) 5 (E) 204

24. Beatrice has many pieces like the gray one in the picture. At least how many of these gray pieces are needed to make a completely full gray square?

(A) 3 (B) 4 (C) 6 (D) 8 (E) 16

Problems from Year 2014

Problems 3 points each

1. Which small figure could be the central part of the larger figure with the star?

       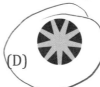 
   (A)　　　　　(B)　　　　　(C)　　　　　(D)　　　　　(E)

2. Jackie wants to place the digit 3 somewhere in the number 2014. Where should she place the digit 3 to make the resulting five-digit number as small as possible?

   (A) in front of 2014　　(B) between the 2 and the 0　　(C) between the 0 and the 1
   (D) between the 1 and the 4　　(E) after 2014

3. Which houses are made using exactly the same triangular and rectangular pieces?

   (A) 1, 4　　　(B) 3, 4　　　(C) 1, 4, 5　　　(D) 3, 4, 5　　　(E) 1, 2, 4, 5

4. When Koko the Koala is not sleeping, he eats 50 grams of leaves per hour. Yesterday he slept 20 hours. How many grams of leaves did he eat yesterday?

   (A) 0　　　(B) 50　　　(C) 100　　　(D) 200　　　(E) 400

5. Maria performs the subtractions shown to the right and gets the numbers from zero to five as results. She connects the dots, starting at the dot with the result 0 and ending at the dot with the result 5, in increasing order. Which figure does she draw as a result?

   (A)　　　(B)　　　(C)　　　(D)　　　(E)

6. Adam built fewer sandcastles than Martin but more than Susan. Lucy built more sandcastles than Adam and more than Martin. Dana built more sandcastles than Martin but fewer than Lucy. Which of the children built the most sandcastles?

   (A) Martin　　(B) Adam　　(C) Susan　　(D) Dana　　(E) Lucy

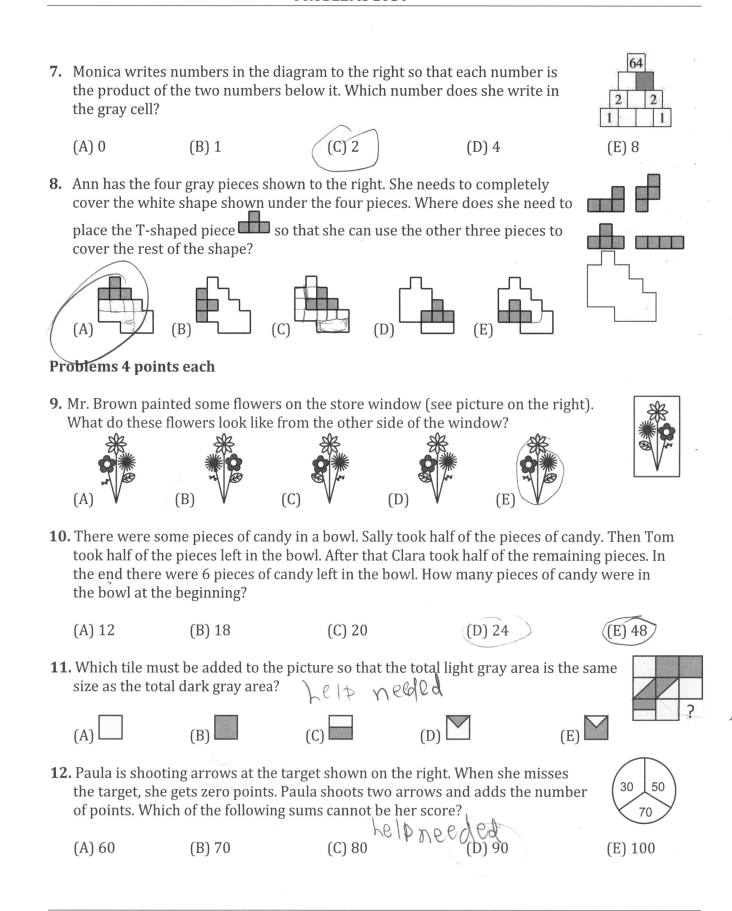

13. Mary had equal numbers of red, yellow, and green tokens. She used some of these tokens to make a pile. You can see all the tokens she used in the figure. After making the pile, she still has five tokens which were not used. How many yellow tokens did she have at the beginning?

(A) 5   (B) 6   (C) 7   (D) 15   (E) 18

14. Peter Rabbit likes to eat cabbage and carrots. In one day, he can eat only 9 carrots, only 2 cabbages, or 1 cabbage and 4 carrots. During one week, Peter ate 30 carrots. How many cabbages did he eat during that week?

(A) 6   (B) 7   (C) 8   (D) 9   (E) 10

15. The solid in the picture to the right was made by gluing eight identical cubes together. What does this solid look like from directly above?

 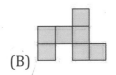   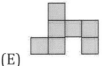
(A)   (B)   (C)   (D)   (E)

16. How many dots are there in this picture to the right?

(A) 180   (B) 181   (C) 182   (D) 183   (E) 265

Problems 5 points each

17. On planet Kangaroo, each kangyear has 20 kangmonths and each kangmonth has 6 kangweeks. How many kangweeks are there in one quarter of a kangyear?

(A) 9   (B) 30   (C) 60   (D) 90   (E) 120

18. Seven children are standing in a circle. No two boys are standing next to each other. No three girls are standing next to each other. Which of the statements below about the number of girls standing in the circle is true?

(A) 3 is the only possible number.   (B) 3 and 4 are the only possible numbers.
(C) 4 is the only possible number.   (D) 4 and 5 are the only possible numbers.
(E) 5 is the only possible number.

19. Eve arranged cards in a line as shown below. In one move, Eve can switch the places of any two cards. What is the smallest number of moves Eve needs to make to get the word KANGAROO?

(A) 2 (B) 3 (C) 4 (D) 5 (E) 6

20. We are making a sequence of triangles out of diamonds. The first three steps are shown. In each step a line of diamonds is added to the bottom. In the bottom line the two outside diamonds, one on each side, are white. All the other diamonds in the triangle are black. How many black diamonds will the figure have in step 6?

(A) 19 (B) 21 (C) 26 (D) 28 (E) 34

21. Kangaroo Hamish bought some of the toys shown in the picture and gave the cashier 150 Kangcoins. He received 20 Kangcoins back. Then he changed his mind and exchanged one of the toys for another. He got back an additional 5 Kangcoins. What toys did Hamish leave the store with?

(A) the bear and the horse
(B) the bear and the train
(C) the bear and the duck
(D) the ball and the duck
(E) the train, the ball, and the duck

22. Write each of the numbers 0, 1, 2, 3, 4, 5, and 6 in the squares to make the addition on the right correct. Which digit will be in the gray square?

(A) 2 (B) 3 (C) 4 (D) 5 (E) 6

23. What is the largest number of small squares that can be shaded in the figure on the left so that no square like the one shown on the right, made of four small shaded squares, appears in the figure?

(A) 18 (B) 19 (C) 20 (D) 21 (E) 22

24. Nick wrote each of the numbers from 1 to 9 in the cells of the 3 × 3 table shown to the right. Only four of the numbers can be seen in the figure. Nick noticed that for the number 5, the sum of the numbers in the neighboring cells is equal to 13 (neighboring cells are cells that share a side). He noticed that the same is also true for the number 6. Which number did Nick write in the shaded cell?

(A) 5   (B) 6   (C) 7   (D) 8   (E) 9

Problems from Year 2015

Problems 3 points each

1.

(A) 6   (B) 7   (C) 8   (D) 10   (E) 15

2. Eric has 10 identical metal strips.

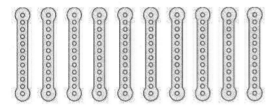

He used screws to connect pairs of them together into five long strips.

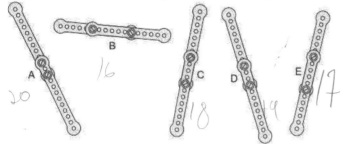

Which strip is the longest?

(A) A   (B) B   (C) C   (D) D   (E) E

3. Which number is hidden behind the square in the equation to the right?

(A) 2   (B) 3   (C) 4   (D) 5   (E) 6

4. Which of the expressions below has the greatest value?

   (A) (1000 − 100) ÷ 10   (B) (1000 − 10) ÷ 9   (C) (1000 − 1) ÷ 9
   (D) (1000 − 100) ÷ 9   (E) (1000 − 10) ÷ 10

5. We start drawing segments connecting every other dot on the circle until we are back at the number 1. The first two segments are already drawn, as shown in the picture to the right. Which figure do we get?

   (A) 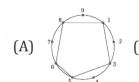 (B)   (C)  (D) 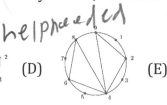 (E)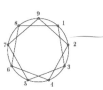

6. A certain whole number has two digits. The product of the digits of this number is 15. The sum of the digits of this number is:

   (A) 2   (B) 4   (C) 6   (D) 7   (E) 8

7. In the picture, we see an island with a highly indented coastline and several frogs. How many of these frogs are sitting on the island?

   (A) 5   (B) 6   (C) 7   (D) 8   (E) 9

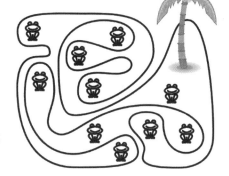

8. This year, March 19 falls on a Thursday. What day of the week will it be in 30 days?

   (A) Wednesday   (B) Thursday   (C) Friday   (D) Saturday   (E) Sunday

Problems 4 points each

9. My umbrella has KANGAROO written on top. It is shown in the picture on the right. Which of the pictures below also shows my umbrella?

   (A)  (B)   (C)   (D)  (E)

© Math Kangaroo in USA, NFP   76   www.mathkangaroo.org

10. Basil wants to cut the shape shown in Figure 1 into identical triangles as shown in Figure 2. How many triangles will he get?

(A) 8   (B) 12   (C) 14   (D) 15   (E) 16

11. Luis had 7 apples and 2 bananas. He gave 2 apples to Yuri, who in return gave some bananas to Luis. Now Luis has as many apples as bananas. How many bananas did Yuri give to Luis?

(A) 2   (B) 3   (C) 4   (D) 5   (E) 7

12. Grandma bought some candy. She gave each of her grandchildren 4 pieces of candy and had 2 pieces left. If she wanted to give each of them 5 pieces of candy, she would be 2 pieces short. How many grandchildren does she have?

(A) 3   (B) 4   (C) 5   (D) 6   (E) 7

13. In a speed skating competition, 10 skaters reached the finish line. The number of skaters who came in before Tom was 3 less than the number of skaters who came in after him. Which place did Tom end up in?

(A) 1   (B) 3   (C) 4   (D) 6   (E) 7

14. Josip has 4 toys: a car, an airplane, a ball, and a ship. He wants to put them all in a row on a shelf. Both the ship and the airplane have to be next to the car. In how many ways can he arrange the toys so that this condition is fulfilled?

(A) 2   (B) 4   (C) 5   (D) 6   (E) 8

15. Pete rides a bicycle in a park that has paths as shown in the picture. He starts from point S and goes in the direction of the arrow. At the first crossroad he turns right, then at the next crossroad he turns left, then right again, then left again, and so on, in that order. Which letter will he not pass?

(A) A   (B) B   (C) C   (D) D   (E) E

16. There are 5 ladybugs (see picture to the right). Two ladybugs are friends with each other only if the numbers of spots that they have differ exactly by 1. On Kangaroo Day each of the ladybugs sent one text greeting to each of her friends. How many text messages were sent?

(A) 2   (B) 4   (C) 6   (D) 8   (E) 9

Problems 5 points each

 17. The figure shown to the left is divided into three identical pieces. What does each of the pieces look like?

 (A)   (B)   (C)   (D)   (E)

18. Jack built a cube using 27 small cubes which are colored either gray or white (see figure). No two of the small cubes which are the same color have a common face. How many white cubes did Jack use?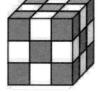

(A) 9   (B) 12   (C) 13   (D) 14   (E) 17

19. We can fill a certain barrel with water if we use water from 6 small pitchers, 3 medium pitchers, and one large pitcher, or from 2 small pitchers, 1 medium pitcher, and 3 large pitchers. If we use only large pitchers of water, how many of them do we need to fill the barrel?

(A) 4   (B) 5   (C) 6   (D) 7   (E) 8

20. The numbers 2, 3, 5, 6, and 7 are written in the squares of the cross (see figure to the right) in such a way that the sum of the numbers in the row is equal to the sum of the numbers in the column. Which of the numbers can be written in the center square of the cross?

(A) only 3   (B) only 5   (C) only 7   (D) 5 or 7   (E) 3, 5, or 7

21. Peter has ten balls, numbered from 0 to 9. He gave four of the balls to George and three to Ann. Then each of the three friends multiplied the numbers on their balls. As the result, Peter got 0, George got 72, and Ann got 90. What is the sum of the numbers on the balls that Peter kept for himself?

(A) 11   (B) 12   (C) 13   (D) 14   (E) 15

22. Three ropes are laid down on the floor as shown on the right. You can make one big, complete rope by adding one of the sets of rope ends shown in the pictures below (without changing their positions). Which of the sets will make one complete rope?

(A)   (B)   (C)   (D)   (E)

23. We have three transparent sheets with the patterns shown to the right. We can rotate the sheets, but not turn them over. Then we put all three sheets exactly one on top of the other. What is the maximum possible number of black squares seen in the square obtained in this way if we look at it from above?

(A) 5   (B) 6   (C) 7   (D) 8   (E) 9

24. Anna, Berta, Charlie, David, and Elisa were baking cookies on Friday and Saturday. Over the two days, Anna made 24 cookies, Berta 25, Charlie 26, David 27, and Elisa 28. Over the two days, one of them made twice as many cookies as on Friday, one 3 times as many, one 4 times as many, one 5 times as many, and one 6 times as many. Who baked the most cookies on Friday?

(A) Anna   (B) Berta   (C) Charlie   (D) David   (E) Elisa

Problems from Year 2016

Problems 3 points each

1. Amy, Bert, Carl, Doris, and Ernst each rolled two dice and added the number of dots.

   Amy   Bert   Carl   Doris   Ernst

   Who rolled the largest total?

   (A) Amy   (B) Bert   (C) Carl   (D) Doris   (E) Ernst

2. Little Kanga is 7 weeks and 2 days old. In how many days will Little Kanga be 8 weeks old?

   (A) 1   (B) 2   (C) 3   (D) 4   (E) 5

3. Which number should be placed in the box with the question mark?

   (A) 24   (B) 28   (C) 36   (D) 56   (E) 80

PROBLEMS 2016

4. What does Pipo the Clown see when he looks at himself in the mirror?

(A) (B) (C) (D) (E)

5. Geoff goes with his father to a circus. Their seats are numbered 71 and 72. Which way should they go?

(A)   (B)   (C)   (D)   (E)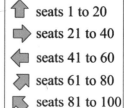

↑ seats 1 to 20
➡ seats 21 to 40
⬅ seats 41 to 60
↗ seats 61 to 80
↙ seats 81 to 100

6. Anna shares some apples between herself and 5 friends. Everyone gets half of an apple. How many apples does she share?

(A) 2 and a half (B) 3 (C) 4 (D) 5 (E) 6

7. A rectangle is partly hidden behind a curtain. What shape is the hidden part?

(A) a triangle (B) a square (C) a hexagon (D) a circle (E) a rectangle

8. Which one of the following sentences correctly describes the picture on the left?

(A) There are as many circles as squares.
(B) There are fewer circles than triangles.
(C) There are twice as many circles as triangles.
(D) There are more squares than triangles.
(E) There are two triangles more than circles.

Problems 4 points each

9. The sum of the digits of the year 2016 is equal to 9. What is the next year, after 2016, where the sum of the digits of the year is equal to 9 again?

(A) 2007 (B) 2025 (C) 2034 (D) 2108 (E) 2134

10. The mouse wants to escape from the maze. How many different paths can the mouse take without passing through the same gate more than once?

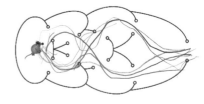

(A) 2 (B) 4 (C) 5 (D) 6 (E) 7

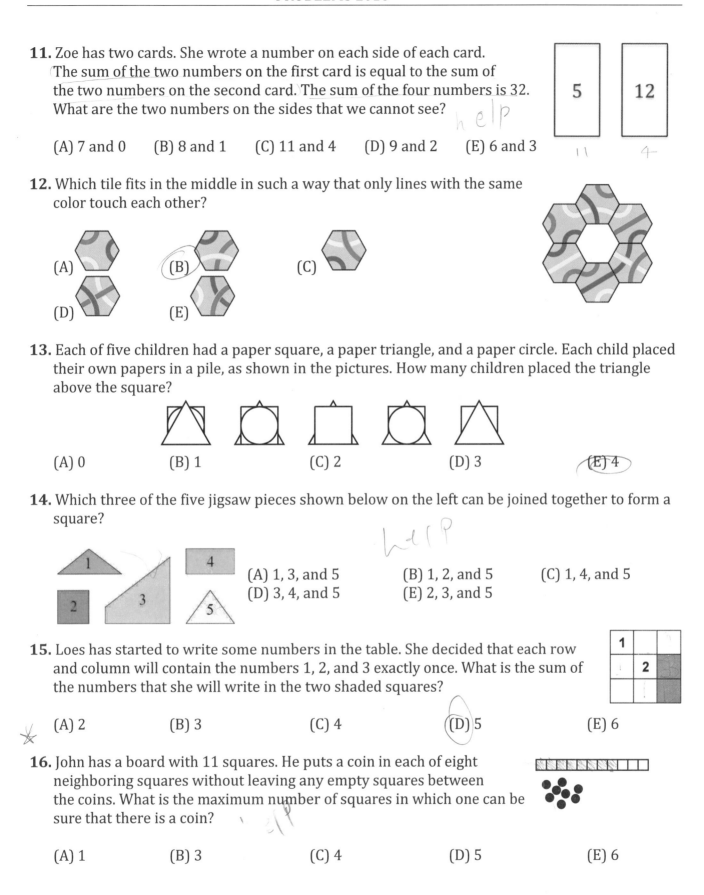

Problems 5 points each

17. When we turn over the card over its right edge, we see the result in the figure to the right. What will we see if we turn this card over its upper edge?

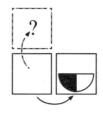

(A)   (B)   (C)   (D)   (E)

18. Tim, Tom, and Jim are triplets (three brothers born on the same day). Their brother Paul is exactly 3 years older. Which of the following numbers can be the sum of the ages of the four brothers?

(A) 25   (B) 27   (C) 29   (D) 30   (E) 60

19. Magic trees grow in a magic garden. Each tree has either 6 pears and 3 apples, or 8 pears and 4 apples. There are 25 apples in the garden. How many pears are there in the garden?

(A) 35   (B) 40   (C) 45   (D) 50   (E) 56

20. My dogs have 18 more legs than noses. How many dogs do I have?

(A) 4   (B) 5   (C) 6   (D) 8   (E) 9

21. Karin wants to place five bowls on a table in order of their weight. She already placed bowls Q, R, S, and T in order. Bowl T weighs the most.

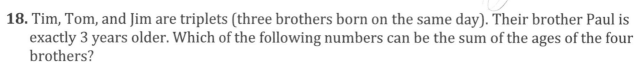

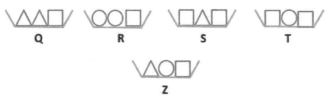

Where must she place bowl Z?

(A) to the left of bowl Q
(B) between bowl Q and bowl R
(C) between bowl R and bowl S
(D) between bowl S and bowl T
(E) to the right of bowl T

22. Rachel adds seven numbers and gets 2016. One of the numbers she adds is 201. She replaces the number 201 with 102. What sum does she get?

(A) 1815   (B) 1914   (C) 1917   (D) 2115   (E) 2118

23. Malte built a bar of 27 bricks.

He breaks the bar into two bars in such a way that one of them is twice the length of the other. Then he takes one of the new bars and breaks it the same way. He continues in this way. Which of the following bars will he not be able to get?

  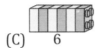 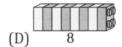 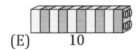

(A) 2 (B) 4 (C) 6 (D) 8 (E) 10

24. Five sparrows are sitting on a branch, as shown in the figure. Each sparrow chirps the same number of times as the number of sparrows it sees. For example, David chirps three times. Then, one sparrow turns to look in the opposite direction. Again, each of the sparrows chirps the same number of times as the number of sparrows it sees. This time, the total number of chirps is more than the first time. Which of the sparrows turned to look in the opposite direction?

Angel   Bertha   Charlie   David   Eglio

(A) Angel   (B) Bertha   (C) Charlie   (D) David   (E) Eglio

Problems from Year 2017

Problems 3 points each

1. Which of the pieces (A) through (E) will fit between the two pieces shown below so the two equations are true?

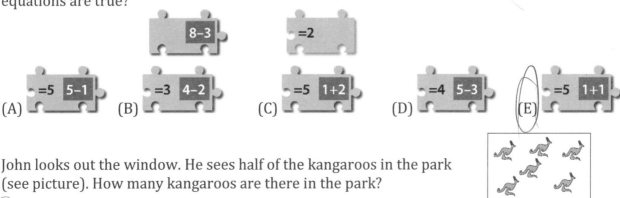

(A)   (B)   (C)   (D)   (E)

2. John looks out the window. He sees half of the kangaroos in the park (see picture). How many kangaroos are there in the park?

(A) 12   (B) 14   (C) 16   (D) 18   (E) 20

3. Two transparent grids have some dark squares, as shown. They both slide into place on top of the board shown in the middle. Now the pictures behind the dark squares cannot be seen. Only one of the pictures can still be seen. Which one is it?

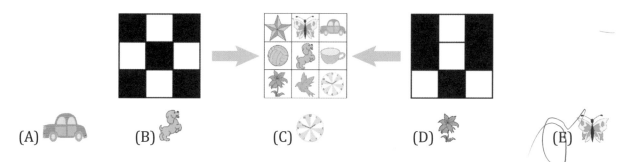

4. A picture of footprints was turned upside down. Which set of footprints is missing?

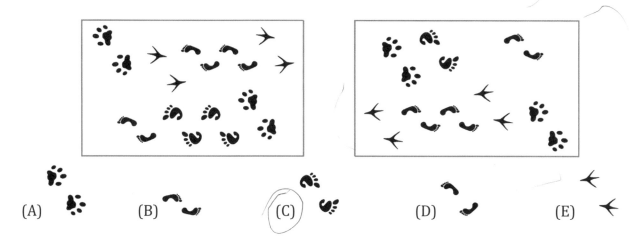

5. What number is hidden behind the panda?

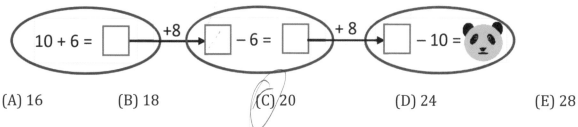

(A) 16   (B) 18   (C) 20   (D) 24   (E) 28

6. The table shows correct sums. What number is in the box with the question mark?

(A) 10   (B) 12   (C) 13   (D) 15   (E) 16

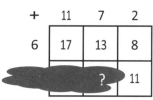

7. Dolly accidentally broke the mirror into pieces. How many pieces have exactly four sides?

(A) 2   (B) 3   (C) 4   (D) 5   (E) 6

8. Here is a necklace with six beads. Which of the pictures below shows the same necklace?

(A)   (B)   (C)   (D)   (E)

Problems 4 points each

9. The picture on the left shows the front of Ann's house. The back of her house has three windows and no door. Which view does Ann see when she looks at the back of her house?

(A)   (B)   (C)   (D)   (E)

10.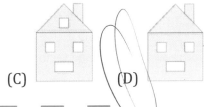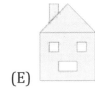

Which of the following is true?

(A) 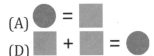   (B) 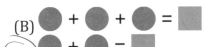   (C)

(D)   (E)

11. Balloons are sold in packets of 5, 10, and 25. Marius buys exactly 70 balloons. What is the smallest number of packets he can buy?

(A) 3   (B) 4   (C) 5   (D) 6   (E) 7

12. Bob folded a piece of paper. He cut exactly one hole in the paper. Then he unfolded the piece of paper and saw the result as shown in the picture to the right. How did Bob fold his piece of paper?

(A)　　　　　　(B)　　　　　　(C)　　　　　　(D)　　　　　　(E)

13. There is a tournament at the pool. First, 13 children signed up, and then another 19 children signed up. Six teams with an equal number of members each are needed for the tournament. At least how many more children need to sign up so that the six teams can be formed?

    (A) 1　　　(B) 2　　　(C) 3　　　(D) 4　　　(E) 5

14. Numbers are placed in the cells of the 4 × 4 square shown in the picture. Mary finds the 2 × 2 square where the sum of the numbers in the four cells is the largest. What is that sum?

    (A) 11　　　(B) 12　　　(C) 13　　　(D) 14　　　(E) 15

15. David wants to prepare a meal with 5 dishes using a stove with only 2 burners. The times needed to cook the 5 dishes are 40 minutes, 15 minutes, 35 minutes, 10 minutes, and 45 minutes. What is the shortest time in which he can do it? (He may only remove a dish from the stove when it is done cooking.)

    (A) 60 minutes　　(B) 70 minutes　　(C) 75 minutes　　(D) 80 minutes　　(E) 85 minutes

16. Which number should be written in the circle with the question mark?

    (A) 10　　　(B) 11　　　(C) 12　　　(D) 13　　　(E) 14

    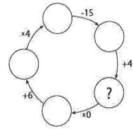

Problems 5 points each

17. The picture shows a group of building blocks and a plan of the same group. Some ink spilled on the plan. What is the sum of the numbers under the ink spills?

    (A) 3　　　(B) 4　　　(C) 5　　　(D) 6　　　(E) 7

    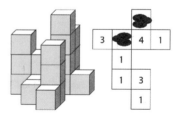

18. How long is the train?

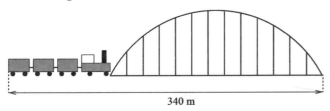
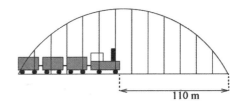

(A) 55 m (B) 115 m (C) 170 m (D) 220 m (E) 230 m

19. The ancient Romans used Roman numerals. We still use them today. I=1, V=5, X=10, L=50, C=100, D=500, M=1000. John was born in February of the year MMVII. How old was John on March 16, 2017?

(A) I (B) III (C) IV (D) V (E) X

20. A small zoo has a giraffe, an elephant, a lion, and a turtle. Susan wants to plan a tour where she sees 2 different animals. She does not want to start with the lion. How many different tours can she plan?

(A) 3 (B) 7 (C) 8 (D) 9 (E) 12

21. Four brothers ate 11 cookies in total. Each of them ate at least one cookie and no two of them ate the same number of cookies. Three of them ate 9 cookies in total and one of them ate exactly 3 cookies. How many cookies did the boy who ate the largest number of cookies eat?

(A) 3 (B) 4 (C) 5 (D) 6 (E) 7

22. Zosia hid a smiley ☺ in some of the cells of the table. In some of the other cells she wrote the number of smileys in the neighboring cells as shown in the picture. Two cells are neighboring if they share a common side or a common corner. How many smileys did she hide?

(A) 4 (B) 5 (C) 7 (D) 8 (E) 11

23. Each of ten bags contains a different number of pieces of candy. The number of pieces of candy in each bag ranges from 1 to 10. Each of five boys took two bags of candy. Alex got 5 pieces of candy, Bob got 7 pieces, Charles got 9 pieces, and Dennis got 15 pieces. How many pieces of candy did Eric get?

(A) 9 (B) 11 (C) 13 (D) 17 (E) 19

24. Kate has 4 flowers, one with 6 petals, one with 7 petals, one with 8 petals, and one with 11 petals. Kate tears off one petal from three flowers. She does this several times, choosing any three flowers each time. She stops when she can no longer tear one petal from three flowers. What is the smallest number of petals which can remain?

(A) 1   (B) 2   (C) 3   (D) 4   (E) 5

Problems from Year 2018

Problems 3 points each

1. Lena has 10 rubber stamps. Each stamp has one of the digits: 0, 1, 2, 3, 4, 5, 6, 7, 8 and 9. She stamps the date of the Kangaroo contest:

   0 3 1 5 2 0 1 8

   How many of the stamps does she use?

   (A) 5   (B) 6   (C) 7   (D) 9   (E) 10

2. The picture shows 3 arrows that are flying and 9 balloons that can't move. When an arrow hits a balloon, the balloon pops, and the arrow keeps flying in the same direction. How many balloons will be hit by the flying arrows?

   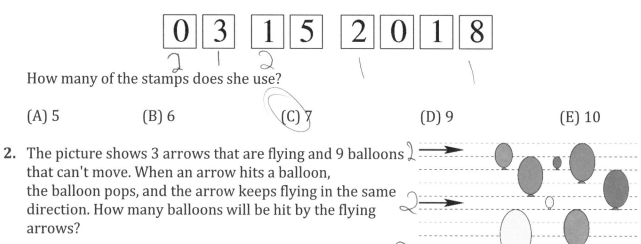

   (A) 2   (B) 3   (C) 4   (D) 5   (E) 6

3. Susan is 6 years old. Her sister is one year younger, and her brother is one year older. What is the sum of the ages of the three siblings?

   (A) 10   (B) 15   (C) 18   (D) 21   (E) 30

4. The picture shows five screws in a block. Four of the screws are the same length. One screw is shorter. Which screw is the short one?

   (A) 1   (B) 2   (C) 3   (D) 4   (E) 5

5. Here is Sophie the ladybug . She turns around. Which of the ladybugs below is not Sophie?

(A) (B) (C) (D) (E)

6. Lucy folds a sheet of paper in half. Then she cuts a piece out of it as shown here: . What will she see when she unfolds the paper?

  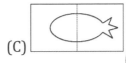 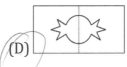 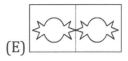
(A) (B) (C) (D) (E)

7. On her first turn, Diana scored 12 points total with three arrows. On her second turn she scored 15 points. How many points did she score on her third turn?

(A) 18 (B) 19 (C) 20 (D) 21 (E) 22

12 points   15 points   ???

8. Mike sets the table for 8 people. He must set the table correctly for each person sitting at the table. Setting the table correctly means that the fork is on the left of the plate and the knife is on the right of the plate. For how many people did Mike set the table correctly?

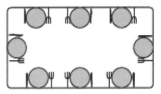

(A) 5 (B) 4 (C) 6 (D) 2 (E) 3

Problems 4 points each

9. Roberto makes designs using tiles like this: . How many of the 5 designs can he make?

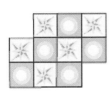 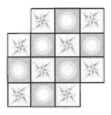 

(A) 1 (B) 2 (C) 3 (D) 4 (E) 5

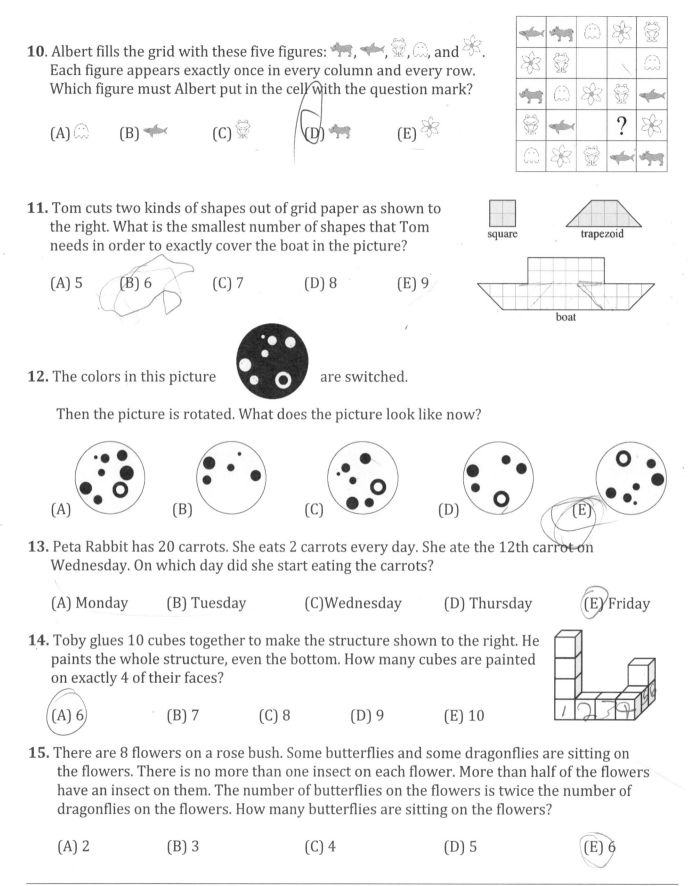

16. Captain Kook wants to sail from the island called Easter through every island on the map and back to Easter. The total journey is 100 kilometers (km) long. The direct distance between Desert and Lake is the same as the distance between Easter and Flower via Volcano. How far is it directly from Easter to Lake?

(A) 17 km   (B) 23 km   (C) 26 km   (D) 33 km   (E) 35 km

Problems 5 points each

17. The rooms in Kanga's house are numbered. Baby Roo enters the main door, passes through some rooms and leaves the house. The numbers on the rooms that he visits are always increasing. Through which door does he leave the house?

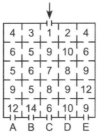

(A) A   (B) B   (C) C   (D) D   (E) E

18. Four balls each weigh 10 g, 20 g, 30 g, and 40 g. Which ball weighs 30 g?

(A) A   (B) B   (C) C   (D) D
(E) It could be A or B.

19. The band shown in the drawing can be fastened in five ways. How much longer is the band fastened in one hole than the band fastened in all five holes?

Unfastened band   Band fastened in one hole

(A) 4 cm   (B) 8 cm   (C) 10 cm   (D) 16 cm   (E) 20 cm

20. In an ancient language the symbols 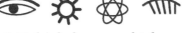 represent the following numbers: 1, 2, 3, 4, and 5. Nobody knows which symbol represents which number.
We know that:

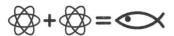      

Which symbol represents the number 3?

(A)   (B)   (C)   (D)   (E)

21. The hexagonal stained-glass tile is flipped. One of the flips is shown. What does the stained-glass tile look like at the far right?

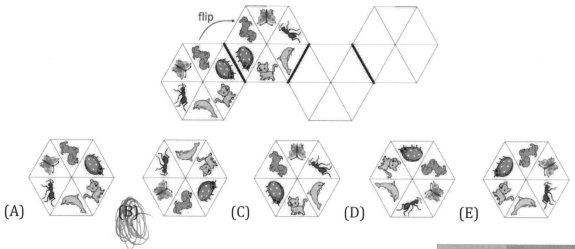

(A)　　　(B)　　　(C)　　　(D)　　　(E)

22. The large rectangle is made up of a number of squares of various sizes. The 3 small squares each have an area of 1. What is the area of the large rectangle?

(A) 165　　(B) 176　　(C) 187　　(D) 198　　(E) 200

23. Leon wants to write the numbers from 1 to 7 in the grid shown. Two consecutive numbers cannot be written in two neighboring cells. Neighboring cells are those that meet at the edge or at a corner. What numbers can he write in the cell marked with the question mark?

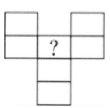

(A) all seven numbers　　(B) all of the odd numbers
(C) all of the even numbers　　(D) only the number 4
(E) only the numbers 1 or 7

24. To defeat a dragon, Matthias has to cut off all the dragon's heads. If he can cut off 3 of the dragon's heads, one new head immediately grows. Matthias defeats the dragon by cutting off 13 heads in total. How many heads did the dragon have at the beginning?

(A) 8　　(B) 9　　(C) 10　　(D) 11　　(E) 12

Problems from Year 2019

Problems 3 points each

1. The higher the step on the podium, the higher the rank of the runner. Who finished third?

   (A) A (B) B (C) C (D) D (E) E

2. In the pictures, each dot stands for 1 and each bar stands for 5. For example, 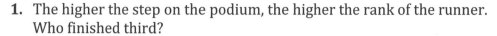 stands for 8. Which picture stands for 12?

   (A)   (B)   (C)   (D)   (E)

3. Yesterday was Sunday. What day is tomorrow?

   (A) Tuesday   (B) Thursday   (C) Wednesday   (D) Monday   (E) Saturday

4. There are two holes in the cover of a book. When the book is open, it looks like this:

   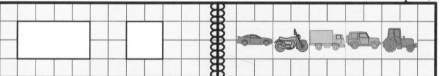

   Which pictures does Olaf see through the holes when he closes the book?

   (A) ,   (B) ,   (C) ,

   (D)  (E)

5. Karina cuts out one piece like this from the sheet shown to the right. Which piece can she get?

   (A)   (B)   (C)   (D)   (E) ♥♥

6. Three people walked across a field of snow wearing muddy shoes. In which order did they do this?

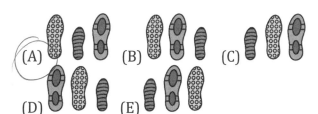
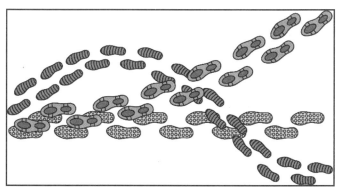

7. Pia is making shapes using the connected sticks shown in the picture.

Which of the following shapes uses more sticks than Pia has?

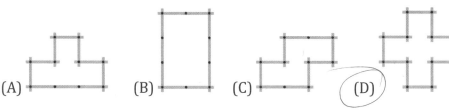

8. What number should replace the question mark when all the calculations are completed correctly?

(A) 4   (B) 5   (C) 6   (D) 7   (E) 8

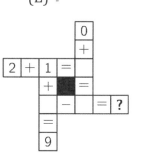

Problems 4 points each

9. Linda pinned up 3 photos in a row on a cork board using 8 pins. Peter wants to pin up 7 photos in the same way. How many pins does he need?

(A) 14   (B) 16   (C) 18   (D) 22   (E) 26

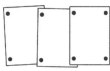

10. Dennis wants to remove one cell from the shape: 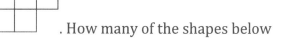 . How many of the shapes below can he get?

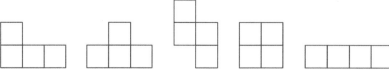

(A) 1   (B) 2   (C) 3   (D) 4   (E) 5

11. Six strips are woven into a pattern as shown: .

What does the pattern look like from the back?

(A)　　(B)　　(C)　　(D)　　(E)

12. The weight of toy dog is a whole number. How much does one toy dog weigh?

(A) 7 kg　　(B) 8 kg　　(C) 9 kg　　(D) 10 kg　　(E) 11 kg

13. Sara has 16 blue marbles. She can trade marbles in two ways: 3 blue marbles for 1 red marble, and 2 red marbles for 5 green marbles. What is the maximum number of green marbles she can get?

(A) 5　　(B) 10　　(C) 13　　(D) 15　　(E) 20

14. Steven wants to write each of the digits 2, 0, 1, and 9 in one of the boxes of the sum: ☐☐☐ + ? . He wants to get the largest possible answer. Which digit can he write instead of the question mark?

(A) Either 0 or 1　　(B) Either 0 or 2　　(C) Only 0　　(D) Only 1　　(E) Only 2

15. A glass full of water weighs 400 grams. An empty glass weighs 100 grams. How many grams does a glass half-full with water weigh?

(A) 150　　(B) 200　　(C) 225　　(D) 250　　(E) 300

400 g　　100 g　　?

16.

Together we cost 5 cents.　　Together we cost 7 cents.　　Together we cost 10 cents.　　How much do we cost together?

(A) 8 cents　　(B) 9 cents　　(C) 10 cents　　(D) 11 cents　　(E) 12 cents

Problems 5 points each

17. Each shape stands for a different number. The sum of the three numbers in each row is shown to the right of the row. Which number does the stand for?

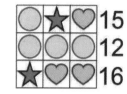

(A) 2 (B) 3 (C) 4 (D) 5 (E) 6

18. Anna used 32 small white squares to frame a 7 by 7 picture. How many of these small white squares does she need to frame a 10 by 10 picture?

(A) 36 (B) 40 (C) 44 (D) 48 (E) 52

19. The pages of a book are numbered 1, 2, 3, 4, 5, and so on. The digit 5 appears exactly 16 times. What is the maximum number of pages this book can have?

(A) 49 (B) 64 (C) 66 (D) 74 (E) 80

20. A hallway has the dimensions shown in the picture. A cat walks on the dashed line along the middle of the hallway. How many meters does the cat walk?

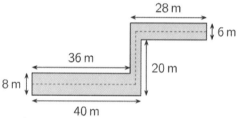

(A) 63 (B) 68 (C) 69 (D) 71 (E) 83

21. In a park there are 15 animals: cows, cats, and kangaroos. We know that precisely 10 are not cows and precisely 8 are not cats. How many kangaroos are there in the park?

(A) 2 (B) 3 (C) 4 (D) 8 (E) 10

22. Mary has 9 small triangles: 3 of them are red (R), 3 are yellow (Y), and 3 are blue (B). She wants to form a big triangle by putting together these 9 small triangles so that any two triangles with an edge in common are different colors. Mary places some small triangles as shown in the picture. Which of the following statements is true after she has finished?

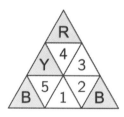

(A) 1 is yellow and 3 is red (B) 1 is blue and 2 is red (C) 1 and 3 are red
(D) 5 is red and 2 is yellow (E) 1 and 3 are yellow

23. There are five children: Alek, Bartek, Czarek, Darek, and Edek. One of them ate a cookie.
   Alek says: "I did not eat the cookie."
   Bartek says: "I ate the cookie."
   Czarek says: "Edek did not eat the cookie."
   Darek says: "I did not eat the cookie."
   Edek says: "Alek ate the cookie."
   Only one child is lying. Who ate the cookie?

(A) Alek   (B) Bartek   (C) Czarek   (D) Darek   (E) Edek

24. Emil started to hang up towels using two pegs for each towel as shown in figure 1. He realized that he would not have enough pegs and began to hang up the towels as shown in figure 2. Altogether, he hung up 35 towels and used 58 pegs. How many towels did Emil hang up in the way shown in figure 1?

(A) 12   (B) 13   (C) 21   (D) 22   (E) 23

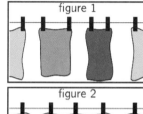
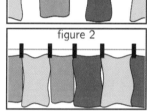

Part II: Solutions

Solutions for Year 1998

1. **(C) 9**
   Each kangaroo had four legs, two ears and one tail, so Bob counted 4 + 2 + 1 = 7 parts for every kangaroo. 63 ÷ 7 = 9, so Bob saw 9 kangaroos. Check:

   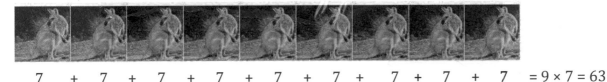

   7 + 7 + 7 + 7 + 7 + 7 + 7 + 7 + 7 = 9 × 7 = 63

2. **(D) 6 × 8 + 20 ÷ (4 − 2)**
   Remember the order of operations. First do what is in the parentheses. Then, from left to right, do multiplication and division, and finally do addition and subtraction, also from left to right.

   (A) 6 × (8 + 20) ÷ 4 − 2 = 6 × 28 ÷ 4 − 2 = 168 ÷ 4 − 2 = 42 − 2 = 40
   (B) (6 × 8 + 20 ÷ 4) − 2 = (48 + 20 ÷ 4) − 2 = (48 + 5) − 2 = 53 − 2 = 51
   (C) (6 × 8 + 20) ÷ 4 − 2 = (48 + 20) ÷ 4 − 2 = 68 ÷ 4 − 2 = 17 − 2 = 15
   (D) 6 × 8 + 20 ÷ (4 − 2) = 6 × 8 + 20 ÷ 2 = 48 + 10 = **58**
   (E) 6 × (8 + 20 ÷ 4) − 2 = 6 × (8 + 5) − 2 = 6 × 13 − 2 = 78 − 2 = 76

3. **(C) 8**
   There are two large overlapping triangles (one marked red and one marked purple in the picture to the right) and six small triangles (marked green).

4. **(D) 7th**
   Apartments 1, 2, 3 are on the 2nd floor; apartments 4, 5, 6 are on the 3rd floor; apartments 7, 8, 9 are on the 4th floor; apartments 10, 11, 12 are on the 5th floor; apartments 13, 14, 15 are on the 6th floor; and apartments 16, **17**, 18 are on the 7th floor. Mary lives on the 7th floor.

5. **(C) 48**
   12 × 12 × 12 = (6 × 2) × 12 × (2 × 6) = 6 × (2 × 12 × 2) × 6 = 6 × **48** × 6.

6. **(C) 4**
   A three-digit number cannot start with a zero. So, the only possible numbers are 307, 370, 703 and 730.

7. **(D) 28**
   In the very center of the tower there are 4 vertical blocks, and around it there are 4 staircase-like structures, each made of 6 blocks. The total number of blocks is 4 + 4 × 6 = 28.

   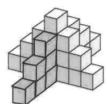

8. (C) 10 minutes
   Try different paths. Zigzagging through the middle part of the figure gives the shortest time.

   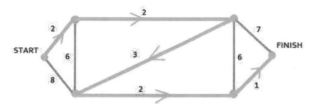

9. (E) 13
   Possible numbers of cookies for 3 plates when 1 cookie is always left over are:
   4, 7, 10, 13, 16, 19, 22, 25, 28, ... (multiples of 3 plus 1).
   Possible numbers of cookies for 4 plates when 1 cookie is always left over are:
   5, 9, 13, 17, 21, 25, 29, ... (multiples of 4 plus 1).
   The smallest number on both lists is 13. If Joanna divides evenly 13 cookies between 2 plates, 1 cookie is also left over.

10. (C) 1290
    To find a solution, reverse your calculations. If something was added, subtract it; if it was subtracted, add it back. 3250 – 2000 = 1250 and 1250 + 40 = 1290.

11. (C) Thursday
    The well is 5 meters deep. On Monday the snail goes up 2 meters but slides 1 meter down. It starts Tuesday with still 4 meters to go up, moves 2 meters up and slides 1 meter down, so on Wednesday it starts with 3 meters left to climb. Again it goes up 2 meters and slides 1 meter down. On Thursday morning it is 2 meters from the top which it climbs during the day and finally gets out of the well.

12. (A)
    Rotate piece Z so it is upside-down, and then piece Z fits with piece (A).

13. (B) 7
    There were 30 runners if we don't count John. They can be divided into 5 groups of equal size: 1 group that finished before him and 4 groups that finished after him. There were 6 runners in each group since 5 × 6 = 30. There was one group of 6 runners before John, so he finished in the 7th place.

14. (D) 24
    The difference between half and one-fourth of a loaf of bread is one-fourth of a loaf of bread. Since-one fourth of a loaf costs 6 pence, the whole loaf costs 4 × 6 pence = 24 pence.

SOLUTIONS 1998

20. (B) 6 years

Look at the line below showing the age differences from youngest to oldest. Each girl is represented by her initial.

1 + 3 + 2 = 6, so Cali is 6 years older than Dorothy.

21. (B) 5

My team received 64 points for ties and wins. The team earned 7 points for 7 ties and 57 points for a number of wins (7 + 57 = 64). The team won 19 games since 19 × 3 = 57 and 7 games were ties, so my team lost 31 − (7 + 19) = 31 − 26 = 5 games.

Solutions for Year 1999

1. (C) 10

Beata's fruits are: three apples, two oranges and five peaches, so Beata has 3 + 2 + 5 = 10 pieces of fruit.

2. (E) 6

The picture on the right has the circles colored to make them easier to see. The part common to all four circles is shaded.

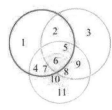

3. (B) 4

To get 5 pieces, you need to break the stick in only 4 places as shown in the picture below.

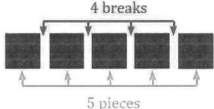

4. (C) 4

Karl is 10 − 3 = 7 years older than Alice. Because the difference of their ages is 7 years, he will be twice her age when she is 7, and he is 14. That will be 14 − 10 = 4 (or 7 − 3 = 4) years from now.

5. (D) Both ways have the same length.

The matching distances are marked by arrows of the same color.

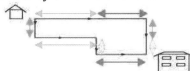

The matching shows that both ways have the same length.

15. (B) 7

First, look at this part of the pyramid:

The pattern shows us to add the two lower numbers together and then divide by 2 to get the top number. In this blank box, we have to write a number such that 5 plus this number is the double of 6. $12 \div 2 = 6$, and $5 + 7 = 12$, so 7 is that number.

Now that we know that there is a 7 in the middle bottom box, let's look at this part:

In the blank box we have to write 8, since $\frac{7+9}{2} = 8$.

At the top of the pyramid we have to write $\frac{6+8}{2}$, which is 7.

Here is the full pyramid.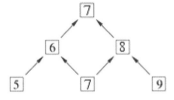

16. (D) 7

If there were 2 or more red balls, then the number of white balls would be at least $7 \times 2 = 14$, and the number of red and white balls would already be more than 15. Thus, there is only 1 red ball and $7 \times 1 = 7$ white balls. $1 + 7 + 7 = 15$, so the other 7 balls are black.

17. (E) 1 dollar and 10 cents

Notice that if Paul added 80 cents to the 30 cents he had left, he would have enough money for another serving of ice cream. So, one serving costs $80 + 30 = 110$ cents, which is equal to 1 dollar and 10 cents.

18. (D) 840

1 meter = 100 cm and 20×5 cm = 100 cm, so the size of the chessboard is 20×20 matches. We will count vertical matches in each row and horizontal matches in each column.

If you have a row of 20 adjacent squares, then the number of vertical matches is $20 + 1 = 21$. There are 20 rows, so the number of all vertical matches is 20×21.
We can count the number of horizontal matches in the same way, so 20×21 is the number of all horizontal matches. The number of all the matches is $2 \times 20 \times 21 = 840$.

19. (B) 15

The numbers, in their natural order, are: 104, 113, 122, 131, 140; 203, 212, 221, 230; 302, 311, 320; 401, 410, and 500. There are 15 numbers on the list.

SOLUTIONS 1999

6. (E) 6
We can split boys into 4 groups, each of the same size as the group of girls. Thus, the class of 30 students consists of 5 equal groups and the number of girls is one fifth of 30, which is 6.

7. (C) 155 g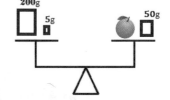
Since the scales must balance, 5 g + 200 g = orange + 50 g
                                 5 g + 150 g + 50 g = orange + 50 g
                                            155 g = orange
The orange weighs 155 g.

8. (B) 24 cm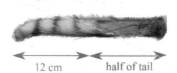
A whole is made up of two halves, so the 12 cm is half
the length the tail. Hence, the tail is 12 cm + 12 cm = 24 cm
long.

9. (E) Saturday
It is also a Sunday 8 weeks after mom's birthday. 8 weeks have 8 × 7 = 56 days and 56 – 1 = 55, so dad's birthday is one day before Sunday, which is Saturday.

10. (B) 7
6 games are not always enough since each team could win exactly 3 games. With 7 games one team must have at least 4 wins and the other team less than 4 wins.

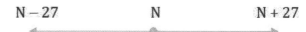

11. (C) 54
Subtracting instead of adding is like moving 27 units in opposite directions from a given number N. Instead moving to the right John moved 27 units to the left. The distance between N – 27 and N + 27 is 27 + 27 = 54, which is the difference between John's result and the result he should have gotten.

12. (A) 64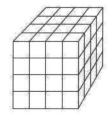
Look at the top layer of the big cube. It is made up of 4 × 4 small cubes since 4 cm = 4 × 1 cm. There are 3 other layers below the top layer, so there are altogether 4 layers of 4 × 4 small cubes. The big cube contains 4 × 4 × 4 = 64 small cubes.

13. (D) 1 kilogram
The pail plus half of the milk weighs 13 kilograms. If we add the other half of the milk to the pail, then the pail is filled with milk to the top and the total weight is 25 kilograms. Hence, the half of the milk weighs 25 kilograms – 13 kilograms = 12 kilograms. All the milk weighs 2 × 12 kilograms = 24 kilograms, so the pail weighs 25 kilograms – 24 kilograms = 1 kilogram.

14. (A) 50 g
Tom eats 5 × 6 g = 30 g of jam from the jar each day. After 20 days he will eat 20 × 30 g = 600 g of jam, so 650 g – 600 g = 50 g of jam will be left in the jar.

15. (D) 1331

The kangaroo has 11 children, and 11 × 11 = (10 + 1) × 11 = 10 × 11 + 11 = 110 + 11 = 121 grandchildren. Each grandchild has 11 children, so the kangaroo has 11 × 121 = (10 + 1) × 121 = 10 × 121 + 121 = 1210 + 121 = 1331 great-grandchildren.

16. (D) 4

Each Kowalski child has a sister and a brother, so there is a girl and a boy in the Kowalski family. The boy has a brother, so there are at least two boys in the family. The girl has a sister, so there are at least two sisters in the family. Thus, the least possible number of children in the family is 4.

17. (D) 110

The page numbers are 10 and 11 since they are consecutive numbers that add up to 21. The product of the two page numbers is 10 × 11 = 110.

18. (C) 4

143 = 40 + 40 + 40 + 23. 3 cows can feed 40 + 40 + 40 children and we need at least one 1 more cow to feed the other 23 children, so Father Virgil needs at least 4 cows.

19. (C) 126

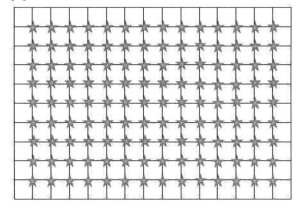

1 m = 10 × 10 cm and 1.5 m = 15 × 10 cm, so the bedspread is a rectangle made out of 10 × 15 square scraps. There are 9 rows of buttons (9 = 10 − 1) and there are 14 buttons (14 = 15 − 1) in each row, so the number of all buttons needed is 9 × 14 = 90 + 36 = 126.

20. (A) 192 cm

The length of Pinocchio's nose is:
2 × 3 cm after the 1st lie, 2 × 2 × 3 cm after the 2nd lie, 2 × 2 × 2 × 3 cm after the 3rd lie,
2 × 2 × 2 × 2 × 3 cm after the 4th lie, 2 × 2 × 2 × 2 × 2 × 3 cm after the 5th lie, and
2 × 2 × 2 × 2 × 2 × 2 × 3 cm after the 6th lie. Consecutive products of 2 are: 2, 4, 8, 16, 32 and 64, so after 6 lies Pinocchio's nose is 64 × 3 cm = 192 cm long.

21. (A) 18

1 pig has as many legs as a duck and a chicken together, so 72 legs (half of the 144 legs) belong to the pigs. Each pig has 4 legs, so there are 18 pigs since 4 × 18 = 72. Hence, there are 18 ducks in the yard since the number of ducks is the same as the number of pigs.

SOLUTIONS 1999

22. (D) 450

The differences are: 501 – 51 = 450, 502 – 52 = 450, 503 – 53 = 450, 504 – 54 = 450, and 505 – 55 = 450. Whichever number was chosen, the difference is always 450.

23. (C) 8

All the grandchildren except one received 10 pieces of candy each. After that Grandma wants each grandchild to have the candy. She can do it if every grandchild has 8 pieces of candy and there are still 6 pieces left over. Every grandchild with 10 pieces of candy must give up 2 pieces of candy, so Grandma can collect 8 pieces for the one grandchild without candy and keep 6 extra pieces. Together she would collect 8 + 6 = 14 pieces, which means that there were 7 grandchildren who originally had 10 pieces each, because 2 × 7 =14. Including the grandchild originally without candy, she has 8 grandchildren.

24. (A)

At 2:00 p.m. the figure looks exactly the same as at 12:00 p.m. since any number of full rotations keeps the original configuration.
Now, we need to see how much it rotates in the 15 minutes after 2 o'clock. Since 15 minutes is ¼ of an hour, the figure will rotate ¼ of a full circle, or 90°, clockwise. The figures below have lines added to help visualize the rotation better. The orange figure in the second picture shows the figure after the rotation.

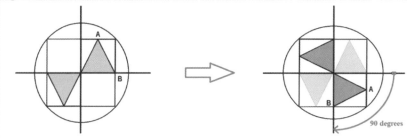

Solutions for Year 2000

1. (B) 15 minutes
   All 10 candles will stay lit for 15 minutes since they are lit at the same time.

2. (C) 40
   Look at the timeline below.

   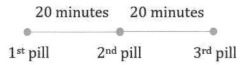

   The little kangaroo will take the last pill
   20 minutes + 20 minutes = 40 minutes
   after taking the first pill.

© Math Kangaroo in USA, NFP   107   www.mathkangaroo.org

3. (D) 222

Number	112	209	312	222	211
Product of digits	2	0	6	8	2
Sum of digits	4	11	6	6	4

Only for 222 the product is greater than the sum.

4. (B) on the 3rd floor
Gavel has to climb one flight of stairs to get to the second floor of the building and Pavel has to climb two flights of stairs: one to get to the second floor, and another one to get from there to the third floor. So, Pavel lives on the 3rd floor.

5. (B) 30 cents
4.50 is four dollars and 50 cents. Each candy bar costs 90 cents, so four candy bars cost 3.60 dollars. Subtracting from the total price of 4.50 dollars, we have 90 cents remaining for the three lollipops, so one lollipop costs 30 cents.

6. (C) 3
160 = 55 + 55 + 50, so at least 3 busses are needed to seat 160 people.

7. (B) 24 minutes

A big square has the area four times greater than the area of a small square. The blue arrows point to matching segments, so the perimeter of the small square is equal to the two sides (marked by green arrows) of the big square. Hence, the perimeter of the big square is twice the perimeter of the small square. Thus, the time to walk around the bigger square plaza is twice the time it takes to walk around the smaller plaza. 12 minutes are needed to walk around the smaller plaza, so 2 × 12 minutes = 24 minutes are needed to walk around the bigger plaza.

8. (C) 9:30
Every bus needs 2 hours to drive from Zakopane to the Krakow airport since 2 hr × 60 km per hour = 120 km, which is the distance from Zakopane to the Krakow airport. To be at the airport at 11:30, the bus had to leave Zakopane no later than 9:30.

9. (B) 4
Kathy eats two bowls of ice cream during the time in which Betty eats three bowls of ice cream. So for each 5 bowls of ice cream the girls eat, Kathy eats 2 bowls. If both girls ate 10 = 2 × 5 bowls of ice cream, then Kathy ate 4 = 2 × 2 bowls of ice cream.

10. (D) 4, 9, 2, 5
To have the smallest possible three-digit number, select the smallest digit from 49215, which is 1, for the hundreds digit. For the tens digit, select the smallest digit from 50 which is 0, so the ones digit is 8. The removed digits are 4, 9, 2, 5.

11. (B) 12
However many apples Aria takes out from her basket is exactly how many apples will remain in Zoe's basket, so 12 apples were left in both baskets altogether.

12. (D) 33

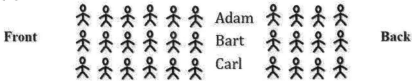

The boys were in the seventh row from the front and the fifth row from the back, so there were six rows before them and four rows behind them. Thus, there are 11 rows with three students in each row. Hence, 11 × 3 = 33 students went to the museum.

13. (B) 4

If 5 (or more) cats were playing moms and each mom had 2 (or more) kids, then the number of all cats would exceed 14 since 5 + 5 × 2 =15. The greatest possible number of cat-moms in the play is 4 as shown below.

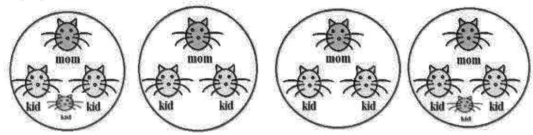

14. (C) 4

According to the second scale, ———— 6 plums, one apple, and one pear can be removed from the first scale, ————, and this scale will stay in balance. Here is the new first scale: ————.

It has two identical groups of plums on the left side and two apples on the right side, so ———— stays in balance. This allows us to remove two plums and one apple from the second scale and gives us the solution: ————.

15. (C) Mr. Jack

Figures (B), (D), and (E) have the same perimeters as (A) as shown by matching arrows (blue is used for horizontal segments and green is used for vertical segments). For (C) vertical matchings are shown but the extra horizontal segments of (C) have no match with horizontal segments of the outer square, so Mr. Jack has the longest fence.

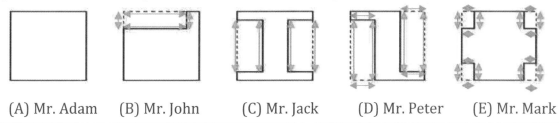

(A) Mr. Adam (B) Mr. John (C) Mr. Jack (D) Mr. Peter (E) Mr. Mark

16. (C) 6

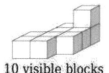
10 visible blocks

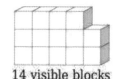
14 visible blocks

The figure on the left weighs 300 grams and consists of 10 blocks, so each block weighs 30 grams. Both figures weigh 900 grams and use identical blocks, so the second figure weighs 600 grams and consists of 20 blocks. Only 14 blocks of the second figure are visible, so 6 blocks are not shown.

17. (B) 12

6 hens eat 8 cups of grain in 3 days, so 3 hens will eat 4 cups of grain in 3 days. Hence, in 9 days, these 3 hens will eat 4 × 3 = 12 cups of grain.

18. (B) 240 cm

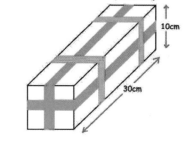

As shown in the picture to the right, there are four pieces of ribbon in all. Two of them (shown in blue), each made of four 10 cm segments, give a total length of 80 cm (2 × 4 × 10 = 80). The other two (shown in orange) have two 10 cm segments and two 30 cm segments each, giving a total length of 160 cm [2 × (2 × 10 + 2 × 30) = 2 × 80 = 160].
This gives a total length of 80 cm + 160 cm = 240 cm.

19. (E) 2

A number line will explain the relationship between ages of the two kangaroos.
The age of the youngest kangaroo is represented by the orange arrow. To multiply it by 5, add 4 green arrows of the same size to it.

The black arrow represents the age of the oldest kangaroo. The difference between the ages is represented by the four green arrows, so it is 4 × the age of the youngest kangaroo. All three kangaroos were born 4 years apart, so the oldest kangaroo is 8 years older than the youngest one. This difference between their ages is always 8 years, so 8 = 4 × the age of the youngest one. Hence, right now, the youngest kangaroo is 2 years old.

20. (C) 6 hours

time of departure   time of return

If we were to turn the clock's dial so that it shows 12:00 on the first clock (new dial marked red), and turn the dial on the second clock the same way, so that the minute hand is at 12 (new dial also marked red), then the second clock would show 6:00 as the time.
Thus, the time difference is 6 hours.

Note: The hour hand and the minute hand on the watch overlap every 1 hour, 5 minutes, and $\frac{5}{11}$ of a minute.

21. (C) 6

If we add the original number and its half, then we get $\frac{3}{2}$ of the original number. If we double the original number, then we get $\frac{4}{2}$ of the original number. The difference between $\frac{4}{2}$ and $\frac{3}{2}$ is $\frac{1}{2}$, so half of the original number is 3. Therefore, the original number is 6.

22. (A) 1

The faces which are touching between the middle and top dice both must show the same number, which cannot be 6, 5, or 4 visible on the top die and cannot be 2 or 3 visible on the middle die, so it must be 1. Since the dice are identical, each die has two opposite faces with 1 and 6 dots. For the middle die, the face opposite the face with 1 dot must have 6 dots, so the top face of the lowest die must also have 6 dots. Therefore, the bottom of the lowest die has 1 dot.

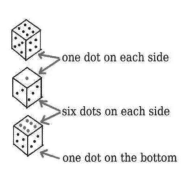

23. (E) There is no such point; this is impossible.

There is only one path connected to point A, so A is either the starting or the ending point of a drawing that is done without lifting the pencil from the paper or going over the same line twice. If A is the starting point, K will need the be ending point, and if A is the ending point, K will have to be the starting point, because there are two other paths that join at K that make the rest of the picture. There is another place where three paths join, which is point C. The only way we can draw three paths coming from one point is if we make it the starting or ending point, so that after coming to the point the first time and moving away from it, we can come back to the point and not have to leave it by a path that has been drawn already (starting at point C means moving away from it, coming back, and moving away again). However, if point C were the starting or ending point, then he could not draw the segment from K to A without going over that line twice. Since this means that the picture cannot be drawn this way, Pete has no starting point to draw the picture of a kangaroo shown without lifting his pencil from the paper and without going over the same line twice.

24. (A) 80 cm

The ball reached a height of 320 cm after the second bounce, so it must have reached a height of 320 cm ÷ 2 = 160 cm after the first bounce, and therefore would have been initially dropped from a height of 160 cm ÷ 2 = 80 cm.

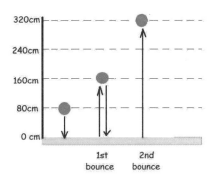

Solutions for Year 2001

1. **(C) 3 hr**
   Each candle burns within 3 hours. Julia lit only 2 candles at first, so there were 2 candles that were not lit at all before the window was closed. Since these two candles were "brand new" when lit after the window was closed, it will take 3 hours for these candles to burn out.

2. **(D) 8**
   Joseph keeps 6 original unbroken sticks and adds 2 new sticks by breaking the seventh stick into two pieces, so he has 8 sticks now.

3. **(D) 400 grams**
   Count all the full squares, (2 + 4 + 5 + 5) × 2 = 32 counted by columns. There are 16 half-squares which is equal to 8 full squares. The area of the heart is 32 + 8 = 40 squares altogether. Thus, the weight of the whole heart is 40 × 10 grams = 400 grams.
   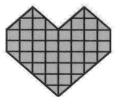

4. **(C) 20**
   Altogether there were 12 × 10 = 120 pairs of shoes in the store. The centipedes bought 3 × 30 + 2 × 5 = 90 + 10 = 100 pairs. Thus, there were 120 – 100 = 20 pairs of shoes left.

5. **(B) 31**
   Kaya gave her mother one cup of berries on the first day. Each day Kaya doubles how many cups of berries she gives her mother, so Kaya gave her mother 1 + 2 + 4 + 8 + 16 = 31 cups of berries over 5 days.

6. **(E) 18 – 6 ÷ 3 = 16**
   Remember to perform the operations in the correct order (multiplication and division first, from left to right, then addition and subtraction from left to right). Also remember that you have to perform the operations within parentheses first.
   Calculations:
   (A) 12 ÷ (4 + 8) = 12 ÷ 12 = 1 ≠ 11
   (B) 8 × 2 + 3 = 16 + 3 = 19 ≠ 40
   (C) 2 × 3 + 4 × 5 = 6 + 20 = 26 ≠ 50
   (D) (10 + 8) ÷ 2 = 18 ÷ 2 = 9 ≠ 14
   (E) 18 – 6 ÷ 3 = 18 – 2 = 16

7. **(E) 5**
   Together there are 19 + 12 = 31 children. The smallest multiple of 6 greater than 31 is 36. 36 – 31 = 5. So, 5 students need to join them.

8. (E) 8 cm
   Altogether, the length of all 4 sticks is 4 × 14 cm = 56 cm. The total space between them is 80 cm – 56 cm = 24 cm. There are 3 gaps between the sticks, so each distance has to be 24 cm ÷ 3 = 8 cm.

9. (C) 3
   Abby will be twice Bobby's age when Bobby will be as old as the difference between their ages, which is 3. This will happen 3 years later, when Abby is 6 and Bobby is 3.

10. (D) 640 m
    The depth of this cave is 221 m + 419 m = 640 m.

11. (E) 5
    The 6 smallest and different numbers are 1, 2, 3, 4, 5, and 6. Their sum is 21 which exceeds 20, so you cannot distribute 20 pieces of candy according to these rules among 6 children. If you ignore the number 1, you can follow the rules and distribute 20 pieces of candy among 5 children; 2, 3, 4, 5, 6 are five different numbers and 2 + 3 + 4 + 5 + 6 = 20. So, the maximum number of children who received candy is 5.

12. (E)
    Each figure is made up of 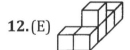 and one more cube. Only in (E) that one cube has a common edge with the corner cube of the piece shown.

13. (C) 50
    By adding 17 and 34 we get 51 cars. The car in which both girls sat in was counted twice, so there were 51 – 1 = 50 cars.

14. (B) 3 times
    Adam and Bart have the same number of chestnuts at the beginning and Adam can split his group of chestnuts into two halves, so Bart can do the same and together they have 4 halves of chestnuts. If Adam gives Bart one half of his chestnuts, then Adam will have 1 half and Bart will have 3 halves, so at the end Bart has 3 times as many chestnuts as Adam has.

15. (C) 3
    Any square has 4 vertices and any triangle has 3 vertices. 17 is neither a multiple of 4 nor a multiple of 3, so there is at least one square and one triangle on the table. Together, one square and one triangle have 7 vertices, so there are still 10 vertices not assigned to any particular figure. 10 is neither a multiple of 4 nor a multiple of 3, so there is yet another pair of a square and a triangle on the table having another 7 vertices altogether. There are still 10 – 7 = 3 unassigned vertices, so the rest of vertices belong to a triangle. Hence, there are 3 triangles on the table.

SOLUTIONS 2001

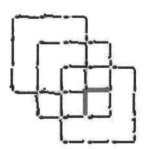

16. (A) 2

There are 8 squares visible in the original picture: 3 are large (3 × 3), 2 are medium (2 × 2) and 3 are small (1 × 1). By adding the two matches shown here in red, 3 more small squares will be added for a total of 11 squares altogether.

17. (C)

The images in the problem have the digit first facing the opposite the direction. Answer (C) shows the number 5 in this way.

18. (C)

The top edge of the folded napkin is one line of symmetry and then, after the first unfolding, the edges to left become one edge which is also a line of symmetry. In terms of the first line of symmetry, the paper cut-out looks like a diamond after the first unfolding and then, after unfolding along the second line of symmetry, we see two diamonds. After rotating the unfolded napkin by 90° we get the napkin shown in (C).

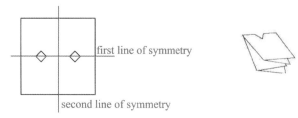

19. (D) 32

At the beginning, there were 12 – 8 = 4 more boys than girls. Since every week one more girl than boy was admitted, it will take 4 weeks for the numbers to be equal. After 4 weeks the number of girls will be 8 + 4 × 2 = 16 and the number of boys will be 12 + 4 × 1 = 16. The club will have 32 members then.

20. (A) 10

All the numbers that fulfill this requirement are (in the decreasing order): 400, 310, 301, 220, 211, 202, 130, 121, 112, and 103. There are 10 such numbers.

21. (A) 600 cm

The length of Anita's or Beth's towel is 720 cm ÷ 4 = 180 cm. Hence the side of the big square is 2 × 180 cm = 360 cm long. To get the width of one of the three smaller towels, divide 360 cm by 3, which is 120 cm. The length of one of these towels is 180 cm (half of the side of the big square). Thus, the perimeter of one rectangular towel is 2 × 120 cm + 2 × 180 cm = 600 cm.

SOLUTIONS 2001

22. (B) $5
Todd has 20 dollars and Will and Kevin together also have 20 dollars (20 + 20 = 40). Since Will has 10 dollars less than Kevin, he has 5 dollars and Kevin has 15 dollars (15 – 10 = 5; 5 + 15 = 20).

23. (D) 6
The table below records the number of candy bars in each basket (L = left, M = middle, R = right) after consecutive actions until the middle basket is empty.

L	M	R	M	L	M	R	M	L	M	R	M	L	M	R	M	L	M	R	M	L	M
10	10	10	9	9	8	9	7	8	6	8	5	7	4	7	3	6	2	6	1	5	0

The right basket contains 6 candy bars and the left basket contains 5 candy bars at the moment when the middle basket is empty, so 6 is the largest number of candy bars left.

24. (E) other number
Remember that the sum of the numbers on the opposite sides of a die is 7. This tells us which numbers are on each of the sides that we cannot see in the picture. So,
after the first move we will have   1 in front, 5 on top, and 3 to the right;
after the second move we will have   2 in front, 1 on top, and 3 to the right;
after the third move we will have   2 in front, 4 on top, and 1 to the right;
after the fourth move we will have   3 in front, 2 on top, and 1 to the right.
The final position is 2 on top which is "other number" from the options listed.

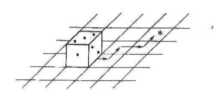

Solutions for Year 2002

1. (B)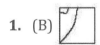

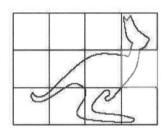

2. (C) 4
2 + 2 – 2 + 2 – 2 + 2 – 2 + 2 – 2 + 2 =
= 2 + 0 + 0 + 0 + 0 + 2 = 2 + 2 = 4

3. (D) 23
Just add the numbers of gifts Andrzej received: 3 + 4 + 3 + 10 + 2 + 1 = 23.

SOLUTIONS 2002

4. (D)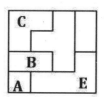
   The picture to the right shows the figures that can be found in the square.

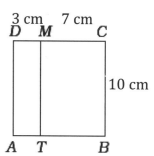

5. (D) Joanna
   Lena and Suzie were born in the same month, so they were born in March. Gina and Suzie were born on the same day of the month, so their birthdays fall on the 20th. This leaves Joanna as the girl who was born on May 17th.

6. (C) 4,200
   There are 60 minutes in an hour, so during one hour the human heart beats on average 60 × 70 = 4200 times.

7. (A) 14 cm
   The lengths of two of the sides of the square ABCD and the longer sides of rectangle ATMD are equal, and the difference in lengths of the other two sides of those figures is 10 cm – 3 cm = 7 cm. The difference between the sum of the lengths of all the sides of the square and the sum of the lengths of all the sides of the rectangle ATMD is 2 × 7 cm = 14 cm.

8. (B) The pictures below show fold lines and edges.

     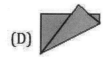 
   (A)   (C)   (D)   (E)

   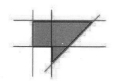 When folding any piece of paper, the result is a shape completely on one side of the folding line, so the red lines (shown to the left) cannot be the folding lines of (B). Green lines potentially could be the folding lines of (B) but unfolding along any of the green lines does not produce a rectangle.

9. (C) 8
   The digit 2 shows up 8 times, in numbers: 2, 12, 20, 21, 22 (2 used twice), 23, and 24.

10. (E) 6
    The scales are balanced when there are 2 melons on the one side and 6 oranges plus 1 melon on the other. By removing one melon from each side, we can see that the weight of 6 oranges is equal to the weight of 1 melon.

11. (C)

    The picture shows where (A), (B), (D) and (E) can be found.

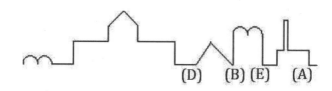

© Math Kangaroo in USA, NFP   116   www.mathkangaroo.org

12. (A) 3

The smallest two-digit number is 10 and the greatest one-digit number is 9. The sum of 10 and 17 is 27. When 27 is divided by 9 the result is 3.

13. (E)

The symbols represent 60 + 60 + 1 + 1 + 1 + 1 = 120 + 4 = 124.

14. (B)

1 + 2 + 3 + 4 + 5 + 6 + 7 + 8 + 9 + 10 + 11 + 12 = 78, so the sum of all the numbers on the face of the clock is 78. The only four consecutive numbers for which the sum is 78 are 18, 19, 20, and 21. In picture (B), the sums of the numbers in each part of the clock are: 12 + 1 + 2 + 3 = 18; 10 + 9 = 19; 11 + 4 + 5 = 20; 6 + 7 + 8 = 21. For each of the other clocks we can find a part with the sum of its numbers less than 18.

15. (C) 15

Klara built a triangle using 6 × 3 = 18 matches. There were 60 – 18 = 42 matches left for Zoe. The length of one of the sides of the rectangle is equal to 6 matches. So, Zoe used 6 + 6 = 12 matches to build that side and the one opposite. The number of matches left for the other two sides is 42 – 12 = 30. So, each of those sides is make out of half of the 30 matches, which is 15.

16. (E) Miki and Niki finished at the same time.

The lengths of the routes taken by the kangaroos are: Miki, 18 units; Niki, 18 units; and Oki, 17 units. Because they were jumping with equal speeds, Miki and Niki finished at the same time because their routes were the same length.

17. (A) Oliver has a dog.

Nate, who has a pet with fur but doesn't like cats, has to have a dog. Because Nate owns a dog, the sentence "Oliver has a dog" is not true. Actually, Oliver owns a cat, and Nate has the dog. The other sentences are true.

18. (B) at 7:20

Mary needs 37 minutes to get to school so Zoe needs 37 – 12 = 25 minutes. Since Zoe arrived at school at 7:45, she had to leave her house 25 minutes earlier, which means she left at 7:20.

19. (D) 18

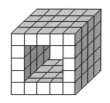 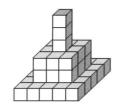

Robert made a tunnel using
5 × 5 × 4 – 3 × 3 × 4 = 100 – 36 = 64 cubes.
This also means that there were 64 cubes left at this point. He made the pyramid out of
5 × 5 + 2 (3 × 3) + 3 = 25 + 18 + 3 = 46 cubes.
Now there were 64 – 46 = 18 cubes left.

SOLUTIONS 2002

20. (E) 11
The green segment below represents the daughter's age. The orange segment represents the period of time after which her mom's future age is 3 times the daughter's future age.

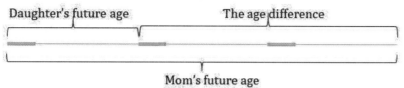

The daughter's future age is represented by the interval with both colors and the mom's future age is represented by 3 such intervals. The difference between the future ages is still 28 since the age difference is always the same. This age difference is twice the daughter's future age. Thus, half of the age difference is 28 ÷ 2 = 14, so the daughter's future age is 14. She is 3 years old now, so the time period from now to this future point in time is 14 − 3 = 11 years.

21. (C) 8
The trio consists of a violinist, a pianist, and a drummer. A violinist alone can be selected in 2 ways. A violinist and a pianist can be selected in 2 × 2 ways since each violinist can be joined by one of 2 pianists. A whole trio is created when a violinist and a pianist are joined by one of 2 drummers, so the conductor can create 2 × 2 × 2 = 8 trios.

22. (D) 21
If 4 medals can be made from 4 plates, then from 16 plates we can cut out 16 medals. The remaining material is enough for 4 more plates. From those 4 plates, 4 medals can be cut out and still the remaining material is enough for 1 more plate. From that 1 plate we can have 1 more medal, therefore there are as many as 16 + 4 + 1 = 21 medals which can be made from 16 plates.

23. (E) 10th
In addition to Tom, there were 27 other students in the math competition. Since there were twice as many students with fewer points than those with more points than Tom, we can divide 27 into 3 groups, which gives 9 students in each group. Two of the groups are the students who got fewer points than Tom, and one group is the students who got more points.
So, there were 9 students who got more points than Tom. Therefore, Tom finished that competition in 10th place.

24. (D) after 21 km
We cannot increase just the ones digit since it is already 9. We have to change the last two digits without using 7 or 8. The smallest such two-digit number is 90, so the odometer reading has to be 187590. To reach it the car has to travel 187590 − 187569 = 21 km.

© Math Kangaroo in USA, NFP   www.mathkangaroo.org

Solutions for Year 2003

1. **(E) 12**
   Follow the shaded line of the grid and count the shaded squares. There are 12 shaded squares.

2. **(C) 4**
   To find the result of 0 + 1 + 2 + 3 + 4 – 3 – 2 – 1 – 0 quickly, notice that all the numbers that are added except for 4 are also subtracted.

3. **(D) 160**
   Since each car has twice as many boxes as the car in front of it, multiply the number in each car by 2 to get the number of boxes in the next car.

   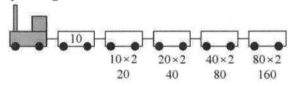

4. **(A) blue**
   blue green **red** yellow
   1   2   **3**   4   - 5 6 **7** 8 - 9 10 **11** 12 - 13 14 **15** 16 - 17

   The pattern consists of four colors repeated in the same order. Because 17 is one more than 16, which is a multiple of 4, we can determine that the 17th color is the same as the first color in the sequence, which is blue.

5. **(C) 50**
   Multiply the number of tables with the same number of chairs by the number of chairs next to the tables. 6 tables with 4 chairs each gives us 6 × 4 = 24 chairs. 4 tables with 2 chairs each gives us 4 × 2 = 8 chairs. 3 tables with 6 chairs each gives us 3 × 6 = 18 chairs. The total number of chairs in the teachers' lounge is 24 + 8 + 18 = 50 chairs.

6. **(D)**
   ♥ ♥ ♥
   ♥ ♥ ♥
   ♥ ♥ ♥
   △ ◇ △

   For every shape that is not a heart there have to be 3 hearts in order for the picture to have 3 times as many hearts as other shapes. Picture D has 9 hearts and 3 other shapes. 3 × 3 = 9.

7. (C) 10

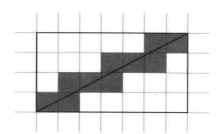

Draw a diagonal as shown and count the squares it passes through. The diagonal passes through 10 squares of the grid.

8. (C) 21 g
There are 9 cubes. Since they are identical, each cube weighs the same amount. The weight of all the cubes is 189 grams or 189 g. The total weight divided by the number of cubes will equal the weight of one cube. 189 g divided by 9 equals to 21 g.

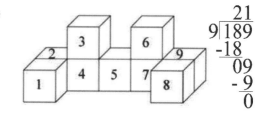

9. (C) 23
Consecutive numbers are numbers in order such as 1, 2, 3, etc. Natural numbers are the normal counting numbers. Philip wrote down seven single-digit numbers from 3 to 9. He still needed to write 35 – 7 = 28 more digits, which means that he wrote fourteen more two-digit numbers, from 10 to 23, so 23 is the greatest number that Philip wrote down.

10. (C) 11 hr 5 min

Anna slept from 9:30 p.m. to 6:45 a.m. as indicated by the shaded part of the clock, so she slept 9 hours and 15 minutes. Since Peter slept 1 hour and 50 minutes longer than Anna, he slept for 11 hours and 5 minutes. Remember that 15 minutes + 50 minutes is 65 minutes, which is 1 hour and 5 minutes.

11. (D) 8
We can describe the pattern as 8 pairs of black-white bars plus the last black bar, so there are 8 white bars in the pattern.

12. (A) 6 m 78 cm
55 dm = 55 × 10 cm = 550 cm and 50 mm = 5 × 10 mm = 5 cm, so Jumping Kangaroo's longest jump during the Olympics was (550 cm + 5 cm) + 123 cm = 678 cm, which is 6 m 78 cm.

13. (C) 18203
Perform the operations in reverse order. From the final result of 20003 subtract 2003, which equals 18000. To get the number which Paul chose at the beginning, add 203 to 18000. The result is 18203.

14. (A) 24

For the hour part, we need to find the greatest sum of the digits on the interval from 00 to 24. For the hours from 20 to 24, the largest sum of digits is 2 + 4 = 6. For the hours from 00 to 19 the largest sum is 1 + 9 = 10, so 10 is the largest sum for the hour part. For the minute part, we need to find the greatest sum of the digits on the interval from 00 to 59, which happens to be 59, with 5 + 9 =14. Barbara will get the greatest sum of the digits on her electronic watch at 19:59 (which is 7:59 p.m.), and that sum is 1 + 9 + 5 + 9 = 24.

15. (B) 24

24 is the number that would double the current number of apples. So, the number of apples Mark actually did pick is also 24.

16. (C) 3 m

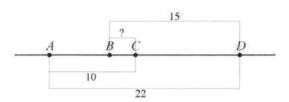

Since the distance from *A* to *D* is 22 m, and the distance from *A* to *C* is 10 m, then the distance from *C* to *D* is 22 m – 10 m = 12 m.
Since the distance from *B* to *D* is 15 m, and the distance from *C* to *D* is 12 m, then the distance from *B* to *C* is 15 m – 12 m = 3 m.

17. (D) 4

Three students do not have siblings, therefore there are 29 – 3 = 26 students who have siblings. The number of sisters and brothers the students have is 12 + 18 = 30 siblings. Because there are more siblings than students, there will be some students who have both a sister and a brother. There are 30 – 26 = 4 students with both a sister and a brother.

18. (E) 9

Once Daniel cuts a piece of paper into three parts, instead of having one piece he now has three. That is two more pieces than he started with, even though they are now smaller.
If we cut 1 piece of paper,   we have 11 + 2 pieces.
If we cut 2 pieces of paper,   we have 11 + 2 + 2 pieces.
If we cut 3 pieces of paper,   we have 11 + 2 + 2 + 2 pieces, and so on.
Thus, 11 + 2 × the number of all pieces cut is 29. Hence, the number of all pieces cut is half of 29 – 11 = 18, which is 9.

19. (E) 1

John bought 3 kinds of cookies: large, medium, and small, so he had to buy at least one of each and pay 4 + 2 + 1 = 7 dollars for these 3 cookies. After that he had 9 dollars left to buy 7 more cookies. 6 cookies of any kind cost at least 6 dollars, so the seventh cookie cannot be a large one since 6 dollars + 4 dollars is 10 dollars (only 9 dollars are left). Therefore, John bought only 1 large cookie. (If we solve the problem further, we can find that he bought 1 large cookie, 3 medium cookies, and 6 small cookies.)

20. (A) 12

Christopher built a rectangular prism with the dimensions of 5 × 4 × 4. To find the number of all the blue inner cubes, we need to remove all the red outer walls (or layers): front and back, top and bottom, left and right sides. Basically, we are reducing the dimensions of Christopher's rectangular prism by 2 units for each dimension. The resulting dimensions are 3 × 2 × 2, so the number of blue inner cubes is 12.

21. (B) 16

Jerry's purchase plans differ by 2 basketballs. To buy those two additional basketballs he needs to borrow 22 dollars and also use the 10 dollars he would have left over had he only bought 5 basketballs. The cost of buying the two extra basketballs is 22 + 10 = 32 dollars. Therefore, one basketball costs 16 dollars.

22. (D)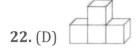

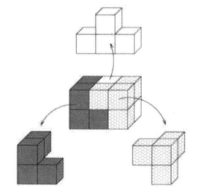

Notice that the bottom of the back wall is made of 3 white cubes and another white cube is visible above its center.

23. (A) 1 and 3

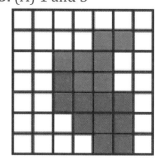

Draw in the lines of the grid over the shaded region. Piece 1 is the only one that can fill in the bottom of the shaded region; it is marked in blue here. Piece 3 needs to be rotated clockwise to fill in the remainder of the shaded region; it is marked in pink.

24. (B) A bear is twice as expensive as a dog.

3 dogs and 2 bears together cost as much as 1 dog and 3 bears since each collection costs as much as 4 kangaroos. Removing 1 dog and 2 bears from each collection results in 2 dogs being priced as much as 1 bear, so a bear is twice as expensive as a dog.

Solutions for Year 2004

1. **(E) 10,015**
   A simple way to find the sum is to notice that 2001+ 2002 + 2003 + 2004 + 2005 =
   = 2000 + 2000 + 2000 + 2000 + 2000 + 1 + 2 +3 + 4 + 5 = 10,000 + 15 = 10,015.

2. **(A) 4 years**
   The difference in age between Mark and his sister is always the same. Today Mark is 9 years old, his sister is 9 – 4 = 5 years old, and the difference between their ages is still 4 years.

3. **(C) 6 km**
   The new straight horizontal path on the bottom is equal to the path *CD*, so the only extra distance to travel are the two paths straight down from *C* and *D*. Thus, the road including the detour is 3 km × 2 = 6 km longer than the original path.

4. **(C) 669 – 391**
   671 – 389 = 282 and the other differences are (in order): 771 – 489 = 282, 681 – 399 = 282, 669 – 391 = 278, 1871 – 1589 = 282, and 600 – 318 = 282.
   Only (C) 669 – 391 = 278 is not identical to the original difference.

5. **(E) 14**
   5 flew away and 3 returned, so there were 2 birds that didn't return. Hence, originally there were 2 more birds than there were at the end. Thus, there were 14 birds at the very beginning.

6. **(B) 1 and 10**
   Numbers 1, 2, 3, 10, 11, and 13 are in the rectangle.
   Numbers 1, 4, 6, 7, 9, 10, 12, and 13, are in the circle. So,
   numbers 1, 10, and 13 are in both the rectangle and the circle.
   However, 13 is also in the triangle, so 1 and 10 are the only numbers
   that are in the rectangle and the circle, but not in the triangle.

   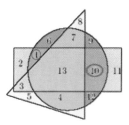

7. **(E) green**
   The red building is next only to the blue building, which means that the red building is either building 1 or 5, and the blue building is either building 2 or 4. The blue building is between the red building and the green building. If the blue building was number 2, then the red building is number 1 and the green building is number 3. If the blue building is number 4, then the red building is number 5 and the green building is number 3. Either way, number 3 is the green building.

8. (B) 3

The figure is made of 24 squares. For the number of shaded squares to equal half the number of white squares, the number of shaded squares has to equal 8 and the number of white squares equal 16. You can get this by dividing 24 into three equal parts, with $\frac{1}{3}$ of 24 being shaded and $\frac{2}{3}$ of 24 not being shaded. There are already 5 squares shaded, so 3 more must be shaded.

9. (D)

To find a match for a completely black rectangle we have to rotate  90° either to the right (clockwise) or to the left (counterclockwise).

The clockwise rotation gives us which is not a match for any of the five options.

Below, see the pictures of the five sheets with the above rectangle placed on top of them.

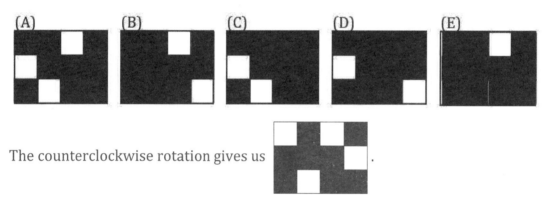

The counterclockwise rotation gives us .

The effect of counterclockwise rotation on the five options is shown below.

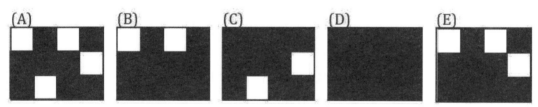

Figure (D) is the match.

10. (D) 9 g

From the first scale, we know that 7 pencils weigh the same as 2 pencils plus 30 grams (g). Remove two pencils from each side of the scale to see that 5 pencils weigh 30 grams, so 1 pencil weighs 6 grams. The second scale shows a pencil plus a pen weighing 15 grams, so the pen weighs 15 grams − 6 grams = 9 grams.

11. (B) 5:05
The four clocks show the times 4:45, 5:05, 5:25, and 5:40.
5:05 is the real time since one clock (the fast one) shows 5:25
and another clock (the slow one) shows 4:45.
The clock showing 5:40 is broken.
There are no other options that have two clocks 20 minutes ahead
and 20 minutes behind of the indicated time.

12. (B) more than half of all the fruit.
Guests ate half of the apples and less than half of the oranges, so they ate less than half of all the fruit. Clearly, more than half of all the fruit remained in the basket.

13. (E) 60,000
Angie divided the original number by 10 and got 600, so the original number was 6,000. Angie should have multiplied the original number by 10 and the correct result should have been 6,000 × 10 = 60,000.

14. (B) 10
Look at the list 1, 2, 3, ..., 23, 24, 25, ..., 43, 44 (the last missing number).
The first 24 numbers are still found in the book, so there are 44 – 24 = 20 missing numbers, which means that there were 20 pages missing (a page is one side of a sheet). Since a sheet has two page numbers, there are 10 sheets missing.

15. (E) Friday
Every seventh day after Eva's birthday is a Tuesday, so 49 days (7 weeks) after her birthday it will be Tuesday again. Three days later, 52 days after Eva's birthday, it will be Friday.

16. (C) 7
The sum of the first row plus the sum of the second row is the sum of the entire table, so the sum of the entire table is 11. The sum of the first column plus the sum of the second column is also the sum of the entire table. The sum of the second column then is 11 – 4 = 7.

17. (E)
When folding each of the figures shown, the top square and the bottom square will touch all four squares in the middle row, so the top square and the bottom square must be the same color. This excludes all figures except (E).
When folding figure (E), the black squares will be opposite, the gray squares will be opposite, and the white squares will be opposite as well.

18. (C) 64

Label the squares as shown in the picture. The bottom side of square A and the bottom side of square B make up the top side of square C, so 16 plus the length of the bottom side of square B equals 40. So, the length of the bottom side of square B is 40 – 16 = 24. Since B is a square, all sides of square B have lengths equal to 24. The length of the left side of square D is equal to the length of the right side of square B plus the length of the right side of square C. So, the length of the left side of square D equals 24 + 40 = 64.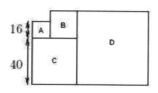

19. (C) 23

Draw a circle, labeled Maria, and draw 16 circles to its right.

By using the last clue, label the 8th circle to Maria's right as Matt.

Count the 8 students already to the left of Matt and add 6 more to get 14 to his left.

Count the students in the picture. There are 23 students in this class.

20. (A) 0

For a 10-digit number to have its digits add up to 9, at least one digit must be 0, because if all digits were at least 1, the sum would be at least 10. Thus, the product of all these digits is 0.

21. (B) 63

Think of the big cube as a stack of five layers. Each layer contains 5 × 5 = 25 cubes.

The top layer has 13 black and 12 white cubes.
The second layer has 12 black and 13 white cubes (even though you cannot see all the cubes, you know that under each cube in the top layer is a cube of opposite color).
The third and fifth layers are the same as the top, and the fourth layer is the same as the second one.
Thus, there are 13 + 12 + 13 + 12 + 13 = 63 white cubes.

22. (B) $375

Peter paid 3 dollars out of 8 dollars, so the boys split the prize into eight equal parts and Peter got 3 of those equal parts. Each equal part is $125, because 1000 ÷ 8 =125, so Peter received $125 × 3 = $375.

23. (E) 3

Focus on the game in which Daniel's team had the one goal scored against them. It could have ended as 0-1, 1-1, 2-1, or 3-1.

In the 0-1 case, the other two games could have ended as 0-0 and 3-0 or 1-0 and 2-0.
In the 1-1 case, the other two games could have ended as 0-0 and 2-0 or 1-0 and 1-0.
In the 2-1 case, the other two games would have ended as 0-0 and 1-0.
In the 3-1 case, the other two games would have ended as 0-0 and 0-0.
Here is the list of all possible results of three soccer games:
  0-1, 0-0, 3-0 which is 1 win, 1 tie, 1 loss   (4 points);
  0-1, 1-0, 2-0 which is 2 wins, 0 ties, 1 loss   (6 points);
  1-1, 0-0, 2-0 which is 1 win, 2 ties, 0 losses (5 points);
  1-1, 1-0, 1-0 which is 2 wins, 1 tie, 0 losses (7 points);
  2-1, 0-0, 1-0 which is 2 wins, 1 tie, 0 losses (7 points);
  3-1, 0-0, 0-0 which is 1 win, 2 ties, 0 losses (5 points).

Daniel's team could not have earned 3 points.

24. (E) M and S

To make gray squares easier to reference, the unknown gray numbers have been assigned letters A through F.

Since each number in the grid is the product of the corresponding gray numbers, $7 \times D = 56$, so D is 8. Also, $6 \times A = 18$, so A is 3. $A \times D = J$, so $J = 24$.

•	A	B	C	7
D	J	K	L	56
E	M	36	8	N
F	T	27	6	P
6	18	R	S	42

$C \times E = 8$ and $C \times F = 6$. Only two integers divide both 8 and 6, which are 2 and 1. So C equals 2 or 1.
If C were 1, then F would be 6 but then there is no integer solution for $27 = F \times B = 6 \times B$.
Therefore, C must be 2. Consequently, F is 3 and E is 4.
$F \times B = 27$, so B is 9.

•	3	B	C	7
8	24	K	L	56
E	M	36	8	N
F	T	27	6	P
6	18	R	S	42

Multiplying the numbers in gray allows us fill in the rest of the chart:

•	3	9	2	7
8	24	72	16	56
4	12	36	8	28
3	9	27	6	21
6	18	54	12	42

•	3	9	2	7
8	24	K	L	56
4	M	36	8	N
3	T	27	6	P
6	18	R	S	42

12 is the only number that occurs twice, and it is found in the cells labeled M and S.

Solutions for Year 2005

1. (C) 500
   $2005 - 205 = 1300 +$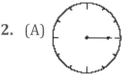

   $2005 - 205 = 1800$
   $1300 + \mathbf{500} = 1800$

2. (A)

   1 quarter of an hour = 15 minutes
   1 hour = 4 quarters of an hour
   17 quarters of an hour = 4 hours and 1 quarter of an hour = 4 hours and 15 minutes
   The time after 4 hours and 15 minutes past noon is 4:15. The minute hand must be one quarter of the way around the face of the clock.

3. (B) 3
   Since Joan got $1 in change, she paid $10 – $1 = $9 for all the cookies.
   Each cookie costs $3 each, so she bought 9 ÷ 3 = 3 cookies.

4. (B) 2
   After the first whistle there were 4 monkeys in each of the 4 rows, so there were 4 × 4 = 16 monkeys at the circus. After the second whistle, the 16 monkeys formed 8 rows, so there were 2 monkeys in each row.

5. (B) 24
   Eva's legs: 2   Parents' legs: 2 × 2 = 4   Brother's legs: 2
   Dog's legs: 4   Cats' legs: 2 × 4 = 8   Parrots' legs: 2 × 2 = 4   Fish legs: 0
   The sum of all the legs is 2 + 4 + 2 + 4 + 8 + 4 + 0 = 24.

6. (D) 60
   The lengths marked in centimeters (cm) also give us the number of rows and columns of pieces of chocolate. The whole bar had 11 × 6 = 66 pieces. The lengths of the sides of the part he ate are 11 – 8 = 3 and 6 – 4 = 2, so he ate 3 × 2 = 6 pieces. Starting with 66 pieces and eating 6 means that there are 66 – 6 = 60 pieces left.

   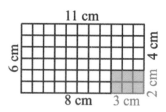

7. (E) a truck that is 325 cm wide and weighs 4250 kg
   Both the width and the weight of the trucks must be equal to or less to the numbers posted on the signs. The width of the truck has to be at most 325 cm AND at the same time the weight has to be at most 4300 kg. Of the trucks listed, only one that is 325 cm wide and weighs 4250 kg is allowed to cross the bridge.

8. (C) 5

The sum of 7 identical numbers must be a multiple of 7. The three-digit number is also a multiple of 10 because it ends in 0. Among the multiples of 10 from 300 to 390 only 350 is a multiple of 7, so the middle digit is 5. Indeed, 50 + 50 + 50 + 50 + 50 + 50 + 50 = 350.

9. (C) 4

In such a family, there have to be at least two boys and two girls. If there was only one girl, she wouldn't have any sisters, and if there was only one boy, he wouldn't have any brothers. There are at least 4 children in this family.

10. (B) 3874

The number 4683 is not even. The number 4874 does not have all different digits. In the number 1246, the hundreds digit in not double the ones digit. Finally, the number 8462 does not have its tens digit greater than its thousands digit. Only the number 3874 satisfies all four conditions. It is even, all the digits are different, 8 is the double of 4, and 7 is greater than 3.

11. (B)

12. (C) 3

Two friends who weigh 80 kg each cannot take the elevator together, because 80 kg + 80 kg = 160 kg, which is more than 150 kg. Only the one friend who weighs 60 kg can take the elevator together with one of the other friends, because 60 kg + 80 kg = 140 kg. The other two friends who weigh 80 kg have to take the elevator separately. So, the four friends need to take at least 3 trips.

13. (D) 45

The difference between the amount of money that Sophia has, compared to Ala and Barb, is the same. Barb has $66 − $24 = $42 more than Ala, so Sophia has half of this difference ($21) more than Ala and $21 less than Barb. Sophia has $24 + $21 = $45.

14. (D) 1

The kangaroo from 3rd cell in 2nd row needs to be moved to the 2nd cell in the 4th row.

15. (E) 8

If the sack didn't have a hole, Greg would have to make just 4 trips. Since he loses half the amount of sand from the sack, the number of his trips needs to be doubled to 8.

16. (D) after 10 days

In 4 days, Adam solved 4 × 5 = 20 problems. Solving 2 problems daily, Brad needed 10 days to solve 20 problems.

17. (B) 3

There are now 15 – 9 = 6 more pieces of paper than before. Cutting one piece of paper into 3 pieces increases the number of all pieces of paper by 2. So, to make 6 new pieces, one needs to cut 3 pieces of paper.

18. (B) 3

Notice that because you're using whole matches, the length and the width need to be whole numbers. Because the perimeter is equal to two times the length plus two times the width, you can just consider what pair of numbers for the length and for the width would add up to 7, which is half of the whole perimeter.
There are 3 possibilities:
 1. A rectangle with sides of 1 match × 6 matches
 2. A rectangle with sides of 2 matches × 5 matches
 3. A rectangle with sides of 3 matches × 4 matches

19. (A) 1 dm

The difference between the outside and the inside perimeters consists of 8 segments (2 at each corner). The length of each one is the same as the width of the frame. Thus, 8 × the width of the frame is 8 decimeters, so the width of the frame is 1 dm.

20. (D) 8

The following steps have to be taken to collect 50 coins:
 Step 1: The trunk needs to be open: **1 lock**
 Step 2: 1st chest needs to be open: **1 lock**
 Step 3: Three boxes in the 1st chest: **3 locks**. At this point, 30 coins can be taken out.
 Step 4: 2nd chest needs to be open: **1 lock**
 Step 5: Two of the boxes in the 2nd chest: **2 locks**. 20 more coins can be taken out.
At least 1 + 1 + 3 + 1 + 2 = 8 locks must be open to take out 50 coins.

21. (C) 24 m

The lengths are indicated by red arrows and the width is indicated by the blue arrow in the picture. Two lengths cover the distance of 16 m, so the length of one flowerbed is 8 m.
The width of one flowerbed is 20 m – 8 m – 8 m = 4 m.
The perimeter of each flowerbed is (2 × 8 m) + (2 × 4 m) = 24 m.

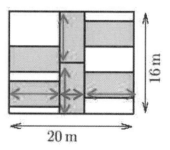

22. (E) 1009

Every two-digit number other than 10 added to 989 gives a four-digit number, so 10 is the only option for the two-digit number. The three-digit number must be 989 + 10, which is 999. The sum of those numbers is 999 + 10 = 1009.

SOLUTIONS 2005

23. (B) 3
None of the cards is in its correct positions, so each of the cards must be moved.
We simply switch two cards in one move. Only one move is needed to place 3 and 4 in their correct position. Just one more additional move will not fix the position of the other three cards, so at least two more moves are needed. 3 moves is the least number of moves needed to place all the five cards in the correct order. One of the switching options is shown below.
   1st move: Switch the cards 3 and 4
   2nd move: Switch the cards 1 and 5
   3rd move: Switch the cards 2 and 5.

24. (E)

Notice that the four small squares will make a checkered pattern on the side opposite the one that is completely black. This excludes the first four options. (E) can happen, since the face opposite the checkered face, which we can't see, could be black.

Solutions for Year 2006

1. (C) 9:00
It takes 6 hours to go up and back from Mount Giewont (3 hours + 30 minutes + 2 hours 30 minutes). The students need to leave at 9:00 a.m. to get back for lunch at 3 p.m.

2. (B) 2006
Remember to do all the multiplication before addition. Any number times 0 is 0.
2 × 0 × 0 × 6 + 2006 = 0 + 2006 = 2006.

3. (D) 7
There are 9 cubes in the top layer of the first figure.
Since there are still 2 cubes left in the top layer of
the second figure, 7 cubes have been removed.

4. (A) Tuesday
Since tomorrow will be Thursday, today is Wednesday. Katie's birthday was yesterday, which was Tuesday.

5. (D) 5
At the beginning John had 10 darts and at the end he had 20, so he gained 20 − 10 = 10 darts. To gain 10 darts, he had to hit the bullseye 10 ÷ 2 = 5 times.

© Math Kangaroo in USA, NFP   131   www.mathkangaroo.org

6. (E) e

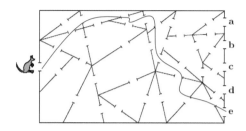

7. (B) 16

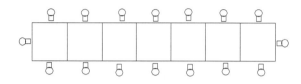

8. (B) $4
$3 = $1 + $2 $6 = $5 + $1 $7 = $5 + $2 $8 = $5 + $2 + $1.
Only $4 cannot be expressed as a sum of $1, $2, $5.

9. (C) 17
The odd-numbered houses are 1, 3, 5, 7, 9, 11, 13, 15, 17, and 19; that is 10 houses.
The even-numbered houses are 2, 4, 6, 8, 10, 12, 14; that is 7 houses.
10 + 7 = 17.

10. (A)

The figure needs to be cut out from a rectangle that is at least 6 squares tall and 6 squares across. Only (A) shows a figure large enough both in height and width.

11. (A) 90
90 = 20 + 10 + 30 + 20 + 10

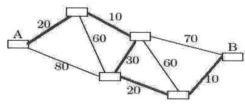

12. (D) 2309415687
The digits on the left side of the number have a greater value based on place value, so we need to start with the smallest digit possible. So, we have to start with the card with number 2. Afterwards, always chose a card from those that remain with the smallest first digit you can find. The remaining cards should be arranged as follow: 309, 41, 5, 68 and 7.

13. (C) 3 and 1

The six weights together weigh 21 pounds. The weights in the first and second box together weigh 17 pounds, so the weights in the third box must weigh 4 pounds. The only option is to have the 1-pound and 3-pound weights in the third box. The first box contains the 4-pound and 5-pound weights, and the second box contains the 2-pound and 6-pound weights.

14. (E) All are equal.

All the routes are equal. All have the length of 8 diagonals of a unit square.

15. (A) 46

8 = 1 × 6 + 2
18 = 3 × 6 + 0
28 = 4 × 6 + 4
38 = 6 × 6 + 2
48 = 8 × 6 + 0
58 = 9 × 6 + 4

The only numbers with the remainder 2 are 8 and 38. These are the numbers on the petals Mary picked out. Their sum is 38 + 8 = 46.

16. (C)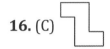

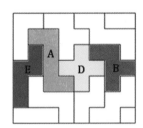

17. (B) 6 feet

The distance between Hanna and Dana is the same as the distance between Dana and Lena, so the distance between Lena and Bennie is the same as the distance between Dana and Lena.

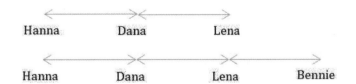

Hence, Lena sits exactly in the middle between Dana and Bennie, so the distances between Hanna and Dana, Dana and Lena, and Lena and Bennie are the same. The distance between Dana and Bennie is twice the distance between Dana and Lena, so the distance between Dana and Lena is half the distance between Dana and Bennie. The distance between Dana and Bennie is 4 feet, so the distance between Dana and Lena is 2 feet. The distance between Hanna and Bennie is 3 times the distance between Dana and Lena which is 6 feet, so Hanna is sitting 6 feet from Bennie.

18. (D) 26

To build the bottom level of a 4-story house from the 3-story house shown to the left, we have to add 3 cards in flat positions (one of them is shown in orange) and below these 3 cards we have to add 4 pairs of cards such as used in the one-story house. Together, 11 new cards are added to the 15 cards of the 3-story house, so Johnny needs 26 cards to build a 4-story house.

19. (D) 36

The surface of the structure consists of top faces of six cubes, left faces of six cubes, and front faces of six cubes. The same is true for bottom, right, and back faces, so Roman painted 6 × 6 = 36 faces of the cubes.

20. (E) Ann and Olga

Kate and Elena live on the same floor since Olga lives on the other floor. Irena and Kate live on the same floor since Ann lives on the other floor. Thus Kate, Elena, and Irena live on the same floor, which must be the second floor. Therefore, Ann and Olga live on the first floor.

21. (B) 2001

If we add or subtract two odd numbers, the result is an even number. By adding or subtracting only even numbers (doesn't matter how many), the result is always an even number.
In our expression we only have two odd numbers, so for any distribution of "+" and "−" the result is always an even number which makes 2001 an impossible result.
The rest of the results can be made as follows:
  1998 = 2002 + 2003 + 2004 − 2005 − 2006
  2002 = 2002 + 2003 − 2004 − 2005 + 2006
  2004 = 2002 − 2003 + 2004 − 2005 + 2006
  2006 = 2002 − 2003 − 2004 + 2005 + 2006.

22. (E) Thursday

March has 31 days. If March has 5 Mondays, then the first Monday of the month cannot happen on March 4th or later. Otherwise, there would not be a fifth Monday in March since 4 + (7 + 7 + 7 + 7) = 32.
There are 3 options for the first Monday of the month, March 1st, March 2nd, or March 3rd.

Monday	1	8	15	22	29
Tuesday	2	9	16	23	30
Wednesday	3	10	17	24	31

Sunday	1	8	15	22	29
Monday	2	9	16	23	30
Tuesday	3	10	17	24	31

Saturday	1	8	15	22	29
Sunday	2	9	16	23	30
Monday	3	10	17	24	31

The tables to the left show other days of the week that also appear 5 times.
Neither Thursday nor Friday appear 5 times in any of these cases, so Thursday is the answer, since Friday is not among the options listed.

SOLUTIONS 2006

23. (C) 4

We have four starting options as shown below. Each of the starting options extends to only one particular solution as shown to the right.

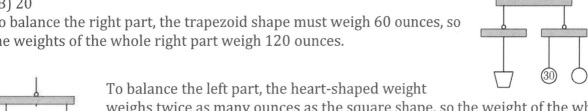

24. (B) 20

To balance the right part, the trapezoid shape must weigh 60 ounces, so the weights of the whole right part weigh 120 ounces.

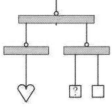

To balance the left part, the heart-shaped weight weighs twice as many ounces as the square shape, so the weight of the whole left part (two hearts and two squares) is equal to the weight of 6 square shape weights (the two squares shown plus two more for each of the hearts). To balance 120 ounces of the right part the square shape must weigh 20 ounces since 6 × 20 ounces is 120 ounces.

A complete graphic solution is shown below.

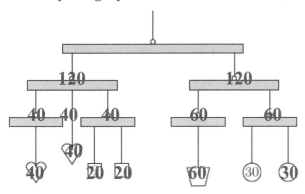

Solutions for Year 2007

1. (C) 2, 3, 5

Anna can't pick up 1 and 2 together, so (A) and (E) are eliminated.
She can't pick up 3 and 4 together, so (B) is eliminated.
She can't pick up 5 and 6 together, so (D) is eliminated.
Along the way Anna could pick up 2, 3 and 5.

2. (A) 1
Only one letter is repeated; it is R.

3. (C)

The relevant rows are row 2, 5, and 6 (in other rows the numbers of squares are identical). Following is a count of squares in those three rows.

Row 2:	3	3	4	3	4
Row 5:	3	3	3	4	3
Row 6:	3	4	4	3	3
Sums:	9	10	11	10	10

The figure (C) consists of the largest number of squares.

4. (C) 3
Five notebooks cost 5 × 80 cents = 400 cents, or $4. Helen has $5, so after buying five notebooks she has $5 − $4 = $1 left to spend on pencils. She can buy at most 3 pencils for 30 cents each since 3 × 30 cents = 90 cents.

5. (C) 64
Between the first and the ninth lamppost, there are eight equal distances of 8 m each.

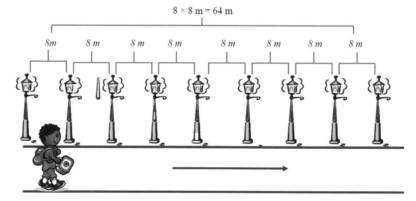

6. (E) 6
Here are all the possible codes:
1 3 5
1 5 3
3 1 5
3 5 1
5 1 3
5 3 1

7. (C) 48

Remember that you need to do multiplication before addition:
4 × 4 + 4 + 4 + 4 + 4 + 4 × 4 = 16 + 4 + 4 + 4 + 4 + 16 = 16 + 16 + 16 = 48

You might also notice that there are four 4's in the middle, which can make up another 4 × 4.
4 × 4 + 4 + 4 + 4 + 4 + 4 × 4 =
4 × 4 +   4 × 4   + 4 × 4 = 3 × (4 × 4) = 3 × 16 = 48

8. (B)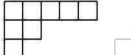

The given figure contains 8 unit squares.
(A) contains 8 unit squares, (B) contains 10 unit squares, (C) contains 9 unit squares,
(D) contains 6 unit squares, and (E) contains 9 unit squares.
Thus, the figure which is given together with any other figure contain 14 to 17 unit squares.
Any rectangle containing the original figure must have one side at least 5 squares long and
the other side at least 3 squares long, so such a rectangle must contain at least 15 = 5 × 3 unit
squares. This eliminates (D) as an option since 8 + 6 = 14 is less than 15.
The given figure and (A) together have 16 unit squares. There are several ways to make a
rectangle which contains 16 square units: 16 = 16 × 1, 16 = 8 × 2, and 16 = 4 × 4. In the first
case 1 is less than 3, in the second case 2 is less than 3, and in the third case both sides are less
than 5, so the original and (A) cannot form a rectangle. The same is true for (C) and (E) since
8 + 9 = 17 and any rectangle containing 17 unit squares must have dimensions 17 × 1. Of
course, 1 is less than 3 which eliminates (C) and (E) as valid options.
(B) is the only option left and we can bring together (B) and the given figure to see a rectangle.
We need to rotate (B) counterclockwise as shown below in blue, and then put the two figures
together.

 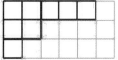

9. (C) 5

To find the number in the shaded cloud, start with the last cloud and reverse the operations.

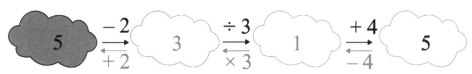

10. (C) 3

Each of the digits 1, 2, and 3 appears in each row and in each column once and only once, so 3 must be the last entry in the first column. The second entry in the last row is also the third entry in the second column (marked in gray), so that entry is neither 3 nor 1. It must be 2, so in the square marked with the question mark Harry can only write 3.
After that there is only one way to fill in the big square, which is shown below.

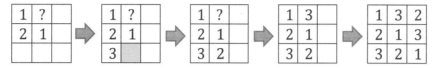

11. (A) 2016

The sum of the digits of 2007 is equal to 9. The next two numbers are 2008 and 2009. For each of those numbers the sum of the digits is greater than 9. After that the tens digit becomes 1, so the next number with the sum of digits equal to 9 is 2016, which is the answer.

12. (C) 17

Annette needs 27 blocks to fill up the whole aquarium since
Length × Width × Depth = 3 × 3 × 3 = 27.
She already put in 10 blocks, the 6 we can see, and 3 more in the bottom layer and 1 in the middle layer that are hidden behind them. 27 – 10 = 17, so she still needs to add 17 cube blocks.

13. (A) January 2nd, 2003

Peter is older than Paul, so he was born first. Add 1 year and 1 day to Peter's birthdate to get Paul's.

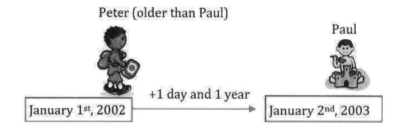

14. (B) 60 m

400 x 15 cm = 6000 cm, and we know that 1 m = 100 cm, so 6000 cm = 60 m.

15. (B) 5

Working backwards, 72 – 19 = 53, which gives us the two digit-number. David wrote the digit on the left (the tens digit) first, so the first digit he wrote was 5.

16. (A) 4 hr 55 min

Rearranging the digital time only using the digits 0, 0, 2 and 7 we could get 02:70, which is not a valid time since the minute part can only be a number from 00 to 59. We could also get 7:02 and 7:20, the earlier of which is 7:02. The difference in time between 07:02 and 02:07 is the difference between 6 hours + 62 minutes and 2 hours + 7 minutes. This difference is 4 hours + 55 minutes, so (A) is the answer.

17. (E) 12

At each edge of the large cube, there is exactly one small cube (the middle one) with exactly two gray sides. Every cube has 12 edges (4 parallel edges in each direction), so there are 12 small cubes that have exactly two gray sides.

18. (B) after 110 km

15951 is the highest palindromic number among numbers of the form 15☐51, so the next palindromic number has the form 16☐61. The smallest among them is 16061 and 16061 − 15951 = 110, so after 110 km the odometer will show a palindromic number again.

19. (C) 65

When we move from one big square to the next one, we add two more rows and two more columns, so the fourth square must have 9 rows and 9 columns. So, the fourth square contains 9 × 9 = 81 small squares.

Remove all the small white squares from all big squares. The first big square is left with 1 row and 1 column of gray squares, the second big square is left with 2 rows and 2 columns of gray squares, and the third big square is left with 3 rows and 3 columns of gray squares. Following this pattern, the fourth big square is left with 4 rows and 4 columns of gray squares. Thus, the fourth big square contains 4 × 4 = 16 gray squares. Therefore, the number of white squares in the fourth big square is 81 − 16 = 65.

20. (A) 1st

Daniel is ahead of Celina and nobody is standing between Bob and Daniel, so Celina is behind Bob and Daniel. Adam is behind Celina, so the four of them are standing in the following order (counting from the farthest away from the cashier): Adam, Celina, Bob, and Daniel. Daniel is not the first in the line, so Eve must be the first one.

21. (A) 48 cm

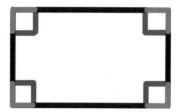

The perimeter of the rectangle is 2 × (15 + 9) = 2 × 24 = 48 cm. The perimeter of the rectangle and the perimeter of the polygon obtained by cutting off four corners of the rectangle share four black segments. The eight red segments of the perimeter of the rectangle match the eight blue segments of the perimeter of the polygon, so the perimeter of the polygon is the same as the perimeter of the rectangle, which is 48 cm.

22. (B) 14

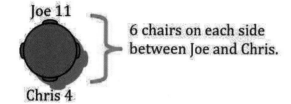

There are 6 chairs (numbered from 5 to 10) on one side of the table between Chris and Joe and the same number of chairs on the other side. 2 × 6 chairs + Chris's chair + Joe's chair = 12 + 1 + 1 = 14 chairs.

23. (D) 192

9 digits are used to write one-digit numbers (1, 2, … , 9). There are 90 two-digit numbers (from 10 to 99), so 2 × 90 = 180 digits are used to write all two-digit numbers. Finally, 3 digits are used to write the number 100. 9 + 180 + 3 = 192 digits.

24. (E) Each of the pictures (A), (B), (C), (D) can represent this unfolded sheet of paper. The sheet of paper is folded twice. Below is another drawing of the folded sheet. Let's make the upper left-hand corner the point where the two fold lines cross. Let's flatten the folded paper to form the smaller square and number the vertices clockwise starting with 1 at the upper left-hand corner. The paper was rotated, so that we do not know which corner was cut. We need to think through the problem taking one corner at a time.

By cutting off corner 1 and unfolding the paper we will see a hole at the center of the paper since the two folding lines intersect at the vertex 1. This is the shape shown by (D).
Vertex 2 is at the two endpoints of the horizontal folding line. If we cut off corner 2 and unfold the paper, then we see shape (C).
Vertex 3 represents the four vertices of the original square sheet of paper. If we cut off corner 3 and unfold the paper, then we see the four vertices of the original square sheet of paper cut off. This is the shape shown by (A).
Vertex 4 is at the two endpoints of the vertical folding line. If we cut off corner 4 and unfold the paper, then we see shape (B).
In conclusion, we can get all four shapes.

SOLUTIONS 2008

Solutions for Year 2008

1. **(C) 21**
   There are 7 days in a week and Ann eats 3 pieces of candy per day, so she eats 7 × 3 = 21 pieces of candy in a week.

2. **(D) $10**
   An adult ticket costs $4 and a child ticket costs $4 − $1 = $3, so one adult and two child tickets cost $4 + $3 + $3 = $10.

   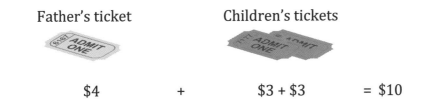

   Father's ticket   Children's tickets

   $4   + $3 + $3   = $10

3. **(B)**  pink roses

   We know that the flowers Luke's aunt and sisters received were of the same color, so these three women received yellow flowers. We also know that Luke's grandmother did not receive roses, so she received red carnations. The bouquet that Luke's mother received were pink roses.

4. **(B) 17**
   37 − 10 = 27, so Adelaide would be left with 27 CDs after giving 10 of them to Mary. At that moment Mary would have 27 CDs and 10 of them are from Adelaide, so she had 17 CDs of her own.

5. **(D) 8**
   1 line passing through the point divides the rectangle into 2 pieces, 2 lines divide the rectangle into 4 pieces, 3 lines divide the rectangle into 6 pieces, and 4 lines divide the rectangle into 8 pieces. Each time you add a new line passing through the point, it increases the number of pieces by 2.

   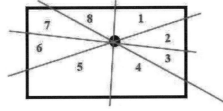

6. (A) 9:30

   6 hours before 4:00 is 10:00 (4 hours back in 12:00, and another 2 hours back makes it 10:00). Going back 30 minutes from 10:00 gives 9:30.

7. (E)

   Charlie can get each of the first four figures, but he can't get the last one. If he places two equilateral triangles in any way he wants, his figure will have at least one angle equal to 60°.

8. (A) 57

   Before the storm, the side of the roof shown in the picture was made of 7 × 10 = 70 tiles and the hole was covered by (counting by rows) 3 + 4 + 4 + 2 = 13 tiles. The number of tiles left is 70 − 13 = 57 tiles.

9. (E) 1988

   The seventh Kangaroo took place 10 years before the seventeenth Kangaroo, so the seventh Kangaroo happened in 1998. Maggie was 10 in 1998, so she was born in 1988.

MATH KANGAROO	1992	1993	1994	1995	1996	1997	1998	1999	2000	2001	2002	2003	2004	2005	2006	2007	2008
	1st	2nd	3rd	4th	5th	6th	7th	8th	9th	10th	11th	12th	13th	14th	15th	16th	17th

10. (D) Jim, Greg, Michael

    The three boys can perform their operations in the following orders:

Greg, Jim, Michael	[(3 × 3) + 2] − 1 = (9 + 2) − 1 = 11 − 1 = 10
Greg, Michael, Jim	[(3 × 3) − 1] + 2 = (9 − 1) + 2 = 8 + 2 = 10
Jim, Greg, Michael	[(3 + 2) × 3] − 1 = (5 × 3) − 1 = 15 − 1 = 14
Jim, Michael, Greg	[(3 + 2) − 1] × 3 = (5 − 1) × 3 = 4 × 3 = 12
Michael, Greg, Jim	[(3 − 1) × 3] + 2 = (2 × 3) + 2 = 6 + 2 = 8
Michael, Jim, Greg	[(3 − 1) + 2] × 3 = (2 + 2) × 3 = 4 × 3 = 12

    There is only one way to end up with 14, so the boys should perform their operation as follows: First Jim adds 2, then Greg multiplies by 3, and finally Michael subtracts 1.

11. (E) Tania

    Grace is taller than Ann and taller than Irena. Irena is taller than Kate, so Grace is taller than Ann, Irena, and Kate. Since Grace is shorter than Tania, Tania is the tallest.

12. (D)

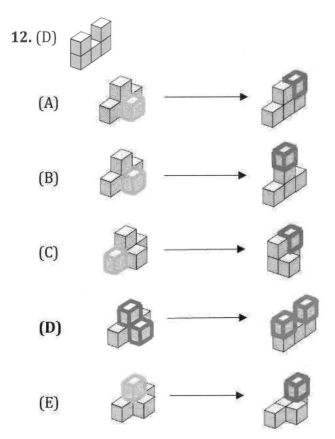

13. (E) 23.
If we take any two different numbers, one is a smaller number and the other is a larger number. In this problem, the sum of the two numbers is 45. The double of the smaller number is less than 45. The double of the larger number is more than 45.
Thus, the smaller number is always less than half of 45 and is at most 22, which means it has to be less than 23.
(The larger number is always more than half of 45, so it is at least 23.)
All possible options of two different numbers with the sum of 45 are shown below.

1 + 44	5 + 40	9 + 36	13 + 32	17 + 28	21 + 24
2 + 43	6 + 39	10 + 35	14 + 31	18 + 27	22 + 23
3 + 42	7 + 38	11 + 34	15 + 30	19 + 26	
4 + 41	8 + 37	12 + 33	16 + 29	20 + 25	

14. (C) 3
5 three-person rooms can hold 5 × 3 = 15 people. 21 – 15 = 6, and the hotel must have 6 ÷ 2 = 3 two-person rooms to hold the 6 people.

15. (B) 29 minutes 3 seconds

```
    6 min   25 sec
   12 min   25 sec
 + 10 min   13 sec
 ─────────────────
   28 min   63 sec = 29 minutes 3 seconds.
```

16. (B) 6

The possible scores are 4, 5, 8, 6, 9:
2 + 2 = 4,
2 + 3 = 3 + 2 = 5,
2 + 6 = 6 + 2 = 8,
3 + 3 = 6,
3 + 6 = 6 + 3 = 9,
6 + 6 = 12.
There are 6 different possible scores.

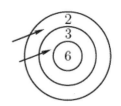

17. (C) 8

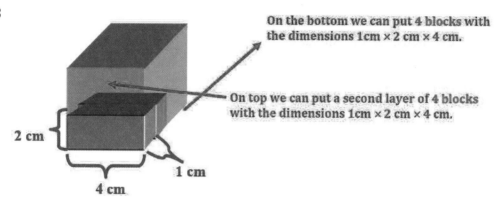

On the bottom we can put 4 blocks with the dimensions 1cm × 2 cm × 4 cm.

On top we can put a second layer of 4 blocks with the dimensions 1cm × 2 cm × 4 cm.

18. (A) 92 kilograms

We can work backwards.

	2004				2005				2006				2007				2008	
	WINTER	SPRING	SUMMER	FALL	WINTER	SPRING	SUMMER	FALL	WINTER	SPRING	SUMMER	FALL	WINTER	SPRING	SUMMER	FALL	WINTER	SPRING
				92 kg		97 kg		93 kg		98 kg		94 kg		99 kg		95 kg		100 kg

− 5 kg   + 4 kg   − 5 kg   + 4 kg   − 5 kg   + 4 kg   − 5 kg

19. (D) 16 m

The perimeter is the distance around a figure. For a square, it is four times the length of one side, because there a square has four sides of equal length. For a rectangle, the perimeter is equal to two times the length of the shorter side added to two times the length of the longer side, because opposite sides are equal.

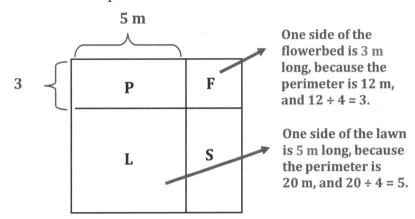

One side of the flowerbed is 3 m long, because the perimeter is 12 m, and 12 ÷ 4 = 3.

One side of the lawn is 5 m long, because the perimeter is 20 m, and 20 ÷ 4 = 5.

The perimeter of the pool is 16 m since 2 × 3 + 2 × 5 = 16.

20. (E) 7

Diane has twice as many brothers as sisters. One of her brothers is Henry. If Diane had only one sister, then she would have to have two brothers. In this case, Henry would have only one brother but two sisters, which can't be correct because Henry has as many brothers as sisters. Suppose that Diane has 2 sisters. Then, she would have to have 4 brothers. If this is the case, then Henry has 3 sisters and 3 brothers, which make the numbers of this brothers and sisters the same. So, there are 3 girls and 4 boys in the family, which means that there are 3 + 4 = 7 children.

21. (E) 36

12, 13, 14, 15, 16, 17, 18, 19,	8
23, 24, 25, 26, 27, 28, 29,	7
34, 35, 36, 37, 38, 39,	6
45, 46, 47, 48, 49,	5
56, 57, 58, 59,	4
67, 68, 69,	3
78, 79,	2
89, +	1
	36

22. (B) 20

The green segment below represents Li's age and the orange segment represents the period of 6 years. The whole segment represents Li's age in 6 years.

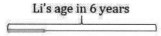

If we double the whole segment, we get Mary's age in 6 years, so Mary's age now is represented by the green-orange-green interval.

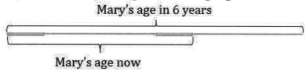

Mary's age is also represented by 5 green segments since Mary is five times as old as Li.

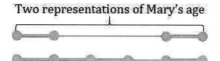

Hence, the orange segment is equivalent to 3 green segments.
The orange segment represents the period of 6 years, so Li's age is 2, Mary's age is 5 × 2 = 10 and in ten years Mary will be 10 + 10 = 20 years old.

23. (C) 8

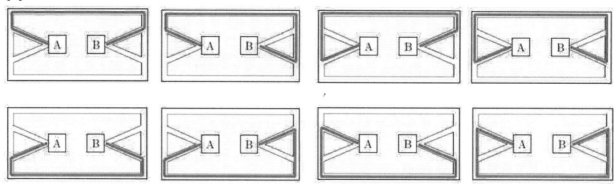

24. (D) 25
Let's reverse the stages of pouring the water back and forth.

At each stage the total amount of water in the two containers is 40 liters.

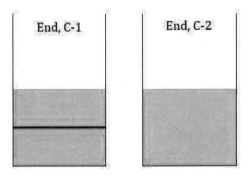

At the end, the total of 40 liters of water is split evenly between two containers. Each container has 20 liters of water and in the first container, C-1, the amount of water was doubled compared to the previous stage, so 10 liters of water were in the first container before this.

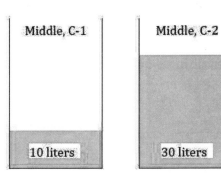

Before the middle stage 5 liters of water were poured from the first container to the second container.

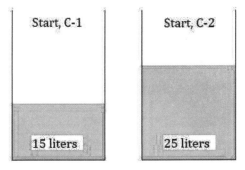

At the beginning, there were 15 liters of water in the first container, and 25 liters of water in the second container.

Solutions for Year 2009

1. (E) 15
There are 3 wooden cubes stacked on top of each other in each of the 5 columns, so 3 × 5 = 15 wooden cubes were used.

2. (C) 2009
Perform the multiplication first, to get 200 × 9 + 200 + 9 = 1800 + 200 + 9 = 2009.

3. (B) In the circle and in the square, but not in the triangle.
The kangaroo is in the region where the circle and square overlap.

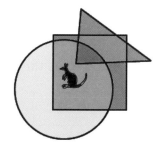

4. (A) 6
Each of the brothers has the same girl as a sister. So, there is only one girl in the family. 5 boys + 1 girl = 6 children in this family.

5. (B) 6
To change the hundreds digit from 9 to 8, only one little square needs to change from white to gray. To change the tens digit from 3 to 0, two little squares need to change from white to gray and one little square from gray to white. To change the ones digit from 0 to 6, one little square needs to change from white to gray and another one from gray to white. There are 6 little squares that need to change color.

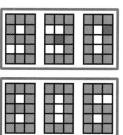

6. (B) 6
Carl ate half of 16 oranges, so he ate 8 oranges. From the remaining 8 oranges, Eva ate 2, so Sophie ate 6 oranges.

7. (C) 46 m
The black zigzag line is made of 5 longer segments and 4 shorter segments.

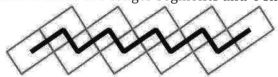

 Each longer black segment connects midpoints of two adjacent tiles in the direction of longer sides, so its length is exactly the length of one tile, which is 6 meters. Each shorter black segment connects midpoints of two adjacent tiles in the direction of shorter sides, so its length is exactly the width of one tile, which is 4 meters. The length of the whole black path is 5 × 6 m + 4 × 4 m = 30 m + 16 m = 46 meters.

8. (D) at 6:53 p.m.
With the commercial breaks, it took 90 + 8 + 5 = 103 minutes for the movie to finish, which is 1 hour and 43 minutes. The movie started at 5:10 p.m., so 1 hour and 43 minutes later means the movie ended at 6:53 p.m.

9. (C) 87 kilograms
The weight difference between the heavier gray kangaroo and the lighter red kangaroo is 35 kilograms. The combined weight of both the gray and the red kangaroos is 139 kilograms. Adding their difference of 35 kilograms to their combined weight is equivalent to twice the weight of the heavier gray kangaroo. 139 kilograms + 35 kilograms = 174 kilograms, so half of it, which is 87 kilograms, is the weight of the gray kangaroo. The lighter red kangaroo weighs 35 kilograms less, which is 52 kilograms.

10. (D) 40

Since Zach first broke off one row of 5 pieces for his brother, the chocolate bar was 5 pieces wide.
After that Zach broke off one row of 7 pieces for his sister, but at this point 1 row of pieces was already gone. So, the whole chocolate bar had 8 rows going across.
There were 5 × 8 = 40 pieces in the whole chocolate bar.

11. (A) 6

There were 25 – 19 = 6 more boys than girls in a dance group to start.
Each week the difference between the number of boys and girls decreased by one, since for every 2 boys there are 3 girls joining.
The number of boys and girls will be the same after 6 weeks.

12. (B) 90

Each cow has 4 legs and each chicken has only 2 legs. For each cow there need to be two chickens for the number of legs to be equal. Since there are 30 cows, there are 2 × 30 = 60 chickens. The farmer has 30 + 60 = 90 animals altogether.

13. (B) 6 cm

One side of the rectangle is 8 cm and the other side is half of 8 cm, which is 4 cm.
The perimeter of the rectangle, or the distance around it, is 2 × 8 cm + 2 × 4 cm = 16 cm + 8 cm = 24 cm. The perimeter of the square is the same as that of the rectangle, and the square has four sides of equal length, so length of one side of the square is 24 cm ÷ 4 = 6 cm.

14. (D) 3

If Magda rolled the die four times and got 6 on each roll she would have obtained 24 points. That would be 1 point more than the 23 points she got, so one of the rolls of the die had to show 5 dots and the other three rolls had to show a 6. There is no other way to get 23 with four die rolls.

15. (B) 2

We need to find three different whole numbers that add up to 7. The middle number will be how many nuts Tola, who did not find the smallest or the greatest number of nuts, found. Trying with the smallest numbers possible, we quickly see that the only numbers that work are 1, 2, and 4. The middle number is 2.

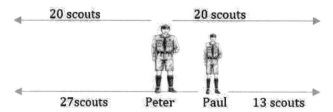

16. (A) 6

Paul had 27 scouts on one side and 13 scouts on the other side of him, so there were 27 + 1 + 13 = 41 scouts standing in a single row. Peter was standing exactly in the middle of the row, so he had 20 scouts on each side of him.

Among the 20 scouts on one side of Peter there are 13 scouts + Paul + the scouts between them, so the number of scouts between Peter and Paul is 6.

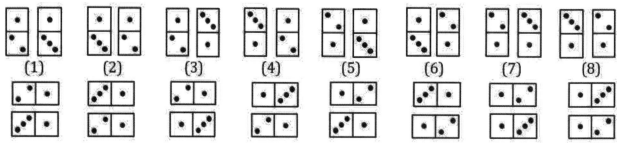

17. (E)

All the possible arrangements of the two dominoes in vertical and horizontal positions are shown below.

Both 2 and 3 dots on the dominoes are always slanted in the same direction, so we cannot get figure (E) as a result.
(A) is a vertical (1), (B) is a vertical (5), (C) is a horizontal (7) and (D) is a horizontal (3).

18. (D) [1][2][*][9][*][8]

In the place of each unknown digit, marked by *, we can put any digits from 0 to 9.
For (A), the sum of the first, third, and fifth digits is 8 + * + 6 = 14 + *, and it can be any number from 14 to 23, depending which digit we chose; adding 0 gives the lowest value and 9 gives the highest. The sum of the second, fourth and sixth digits is 1 + * + 1 = 2 + *, and it can be any number from 2 to 11, so the first sum can never be equal to the second sum.
For (B), the first sum is 7 + 7 + 7 = 21 and the second sum is * + 2 + * = 2 + * + *. The second sum can be any number from 2 to 2 + 9 + 9 = 20, but it is never 21.
For (C), the first sum is 4 + 4 + 4 = 12, and the second sum is * + 1 + 1 = 2 + *. The second sum can be any number from 2 to 11, but it is never 12.
For (D), the first sum is 1 + * + *, and it can be any number from 1 to 1 + 9 + 9 = 19. The second sum is 2 + 9 + 8 = 19. The two sums can be equal.
This happens when the code is [1][2][9][9][9][8].
For (E), the first sum is 1 + 1 + 2 = 4, and the second sum is 8 + * + *. The second sum can be any number from 8 to 8 + 9 + 9 = 26, but it is never 4.
Only the sums in (D) can be equal and only when the code is [1][2][9][9][9][8].

19. (B) 24

The number of eggs Ms. Florentina sold on Tuesday is the difference between the number of eggs she sold on Thursday and the number she sold on Wednesday, so Ms. Florentina sold 96 − 60 = 36 eggs on Tuesday. The number of eggs she sold on Monday is the difference between the number of eggs she sold on Wednesday and the number she sold on Tuesday, so she sold 60 − 36 = 24 eggs on Monday.

20. (D) 4

Here are the possible ways that Maia could have gone:
red → blue → white → yellow but she cannot go red → blue → yellow → white
red → yellow → blue → white but she cannot go red → yellow → white → blue
red → white → blue → yellow
red → white → yellow → blue.

21. (A) 5:00 p.m.

It is 12 hours from 6:15 a.m. to 6:15 p.m., and then another 1 hour and 15 minutes to 7:30 p.m., so Jasper was gone 13 hours and 15 minutes. The clock had been running backwards for the same amount of time, so it went back 12 hours to 6:15 p.m., and then back another 1 hour and 15 minutes to 5:00 p.m.

SOLUTIONS 2009

22. (B) 2

In the original table the sums in the rows are 10, 22, and 10. None of them is a multiple of 3. After any first move there is at least one unchanged row, so the sum of the number in it still is not a multiple of 3. Thus, at least two moves are needed to get the table we want. There are many ways of fixing the table in two moves. Here is a one way: in the first move switch 5 and 7, and in the second move switch 8 and 1 from the third row.

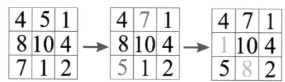

23. (E)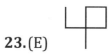

First, let's convert ♥ ♠ ♠ ♠ ♥ ♥ into words. The pattern is: right, left, left, left, right, right. All segments are either vertical or horizontal, so the first segment can be drawn vertically up, vertically down, horizontally right, or horizontally left. After that, the pattern of turns tells us what the figure looks like.

In the pictures below, the black arrow shows the first segment, orange arrows show segments drawn after turning to the right, and blue arrows show segments drawn after turning to the left. In each case, we are following the instructions "right, left, left, left, right, right."

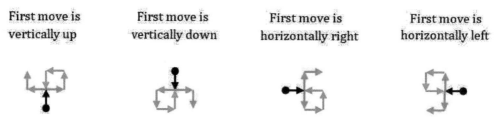

The first figure is exactly the figure in answer (E). The other figures, after rotating, also match the figure (E), so Agnes could have drawn only (E).

24. (B) 5

Among all shoes bought by a group of friends, only the smallest right shoe and only the largest left shoe will be left over if only two shoes are left. All pairs of shoes bought by a group of friends have sizes between 36 and 45. 36 is an even number and 45 is an odd number, so we cannot go from size 36 to 45 by adding only 2 shoe sizes, which would happen if only men were in the group. At least once we have to add 1 shoe size. It indicates that at least one person among the group of friends is a woman. Any two additions of 1 shoe size can be replaced by one addition of 2 shoe sizes, or any two women's shoe sizes can be replaced by one men's shoe size. For the smallest number of people there must be only one woman among the group of friends. The options for the shoe sizes are: 36→37→39→41→43→45, 36→38→39→41→43→45, 36→38→40→41→43→45, 36→38→40→42→43→45 and 36→38→40→42→44→45.

In each case there were 6 pairs of shoes shared among 5 friends.

Solutions for Year 2010

1. (D) 2 − 0 + 1 + 0

   2 + 0 − 1 + 0 = 2 − 1 = 1
   2 − 0 − 1 + 0 = 2 − 1 = 1
   2 + 0 − 1 − 0 = 2 − 1 = 1
   2 − 0 + 1 + 0 = 2 + 1 = 3
   2 − 0 − 1 − 0 = 2 − 1 = 1

2. (C) 12:10

   Since the lesson lasts 40 minutes, half-way through is 20 minutes into the lesson. 20 minutes after 11:50 is 12:10.

3. (C) Daniel's

   The person whose steps are the longest needs to make the least number of steps to cover a given distance, so Daniel's steps were the longest.

4. (C) 31

   The first day, Adam solved 1 problem. The second day, he solved 2 × 1 = 2 problems. The third day, he solved 2 × 2 = 4 problems. The fourth day, he solved 2 × 4 = 8 problems. The fifth day, he solved 2 × 8 = 16 problems. Altogether, he solved 1 + 2 + 4 + 8 + 16 = 31 problems.

5. (D)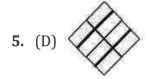

   Notice that in picture (D) the lines run across the small squares and not from corner to corner, like the line does in the picture of the piece of cardboard. All the other pictures can be easily made from the 4 pieces of cardboard.

6. (A) $3

   Buying the items separately would cost $4 + $9 + $5 = $18. So, by buying the set mother would save $18 − $15 = $3.

7. (C) 54

   Work backwards. Before Eva bought the 16 pairs of shoes, she did not have shoes on 7 + 16 = 23 pairs of feet. This means that she had shoes on 50 − 23 = 27 pairs of feet, so she had 2 × 27 = 54 feet with shoes on before she bought the 16 pairs of shoes.

8. (B) 2

9. (E) the number of seconds in one week.
   There are 60 seconds in one minute, 60 minutes in one hour, 24 hours in one day, and 7 days in one week, so 60 × 60 × 24 × 7 is the number of seconds in one week.
   For the other options we have the following expressions:
   (A) 60 × 24 × 7 × 7,   (B) 60 × 60 × 7,   (C) 60 × 24 × 7 × 24,   (D) 24 × 60.

10. (B) Adam, Alex, Tom, Luke
    From the statement, "Tom ate more ice cream than Luke but less than Alex," the order of these three boys from the one who ate the most to the one who ate the least is Alex, Tom, Luke. Since Adam ate more ice cream than Alex, Adam ate the most ice cream.

11. (B) 14
    When Matthew was on the 8th floor, he was already half way through his 12-floor climb, so he still had 6 more floors to go. Hence, Clara lives on 8 + 6 = 14th floor.

12. (D) 20
    Along each vertical edge there are 3 cubes with exactly two sides painted, so along the four vertical edges there are 4 × 3 = 12 such cubes. Along each horizontal bottom edge there are 2 cubes with exactly two sides painted, so along the four bottom edges there are 4 × 2 = 8 such cubes. There are no additional cubes like these along the top horizontal edges.
    Altogether, there are 12 + 8 = 20 cubes with exactly two sides painted.

13. (D) $200
    The perimeter of the fence to be painted is 2 × 20 meters (two longer sides) + 10 meters (one shorter side) which is 50 meters. Using one can of paint Grandpa can paint 10 meters of the fence since it takes one-half of a can to paint 5 meters of the fence. Using five cans of paint Grandpa can paint the whole fence since 50 meters = 5 × 10 meters. One can of paint costs $40, so Grandpa will pay 5 × $40 = $200 for the paint he needs.

14. (C)

The ones digits of the numbers in any row are either   1   2   3   4   5
                                                                           or   6   7   8   9   0.

This excludes (A), (D) and (E) since (A) has 3 as the ones digit in the second column, (D) has 1 and 6 as the ones digits in the second column, and (E) has 7 as the ones digit in the third column. (B) is excluded since the numbers in the table are increasing from one row to the next. (C) represents the 14th and 15th rows of the table.

66	67	68	69	70
71	72	73	74	75

15. (D) 20

At each corner of a little square there two edges labeled with numbers. One of the numbers is odd and the other number is even. In the clockwise direction the order is always the odd number first and then the even number next. Anne always cut along the edges labeled with even numbers, so the sum is 2 + 4 + 6 + 8 = 20.

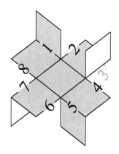

16. (A) 99

Notice that each of the numbers shown in the bottom row of the table is 10 more than the number above it. Since there are 10 numbers shown in the bottom row, the difference between the two rows, not counting the last column, is 10 × 10 = 100. So, to make the sums in both rows equal, the last number must be 100 less than the number above it. The value of ∗ is 199 − 100 = 99.

17. (E)

There is a cut at every place that was a corner when the piece of paper was folded, which means that the four corners of the large square are clipped and there is a hole in the middle.

18. (C) 2003

Since none of the children were right, and Anna's guess of 2010 differs from 1998 by 12 and from 2015 by 5, her guess must differ from the right number by 7 (otherwise, the right answer would have been 1998 or 2015, and we know it was not the case). So, the number of the books was either 2010 − 7 = 2003 or 2010 + 7 = 2017. It cannot be 2017 since 2017−1998 = 19, which is neither 12 nor 5. The other option is 2003. 2003 − 1998 = 5 and 2015 − 2003 = 12, so there are 2003 books in the library.

19. (D) 22

Tom counted 12 – 3 = 9 chairs before he got to where Adam started counting. Adam counted 18 – 5 = 13 chairs before he got to where Tom started counting. There were 9 + 13 = 22 chairs around the table.

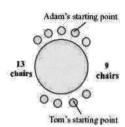

20. (D) 16 mm

The two-link overlap has the width of 2 × 0.5 mm = 1 mm. The first link has the length of 4 mm and each additional link will only increase the length of the chain by 4 mm – 1 mm = 3 mm. A chain made up of five links will be 4 mm + 4 × 3 mm = 4 mm + 12 mm = 16 mm long.

21. (E) 5

At least one red and one yellow ladybug came to the party. Together this pair of ladybugs had 6 + 10 = 16 spots. The other ladybugs at the party had a total of 42 – 16 = 26 spots. 26 is neither a multiple of 6 nor a multiple of 10, so among the other ladybugs there must be another pair of ladybugs of different colors. Together, the two pairs of ladybugs had 2 × 16 = 32 spots, so 42 – 32 = 10 spots were not counted yet. All these 10 spots belong to one yellow ladybug, so 3 yellow ladybugs and 2 red ladybugs came to the party, which is 5 ladybugs altogether.

22. (B) 8

The number of days in any month does not exceed 31. To get the sum of 35, the number of the month needs to be at least 4. Since April is the 4th month and only has 30 days, it cannot be included. Basil's friends had to be born in the 8 different months from May, which is the 5th month, to December, so 8 is greatest number of Basil's friends.

23. (D) Washington

Paul is not from Washington, Pittsburgh, or New York, so he must be from Dallas. Micah is not from Dallas and he is not from Washington or Pittsburgh, so Micah is from New York. Jeff is not from Pittsburgh, so he is from Washington.

24. (E) 96

Olivia's grandmother is 10 × 6 = 60 years old. Olivia's mother is 60 – 14 – 10 = 36 years old. Olivia's great-grandmother is 60 + 36 = 96 years old.

Solutions for Year 2011

1. (A) 20 + 11

   (A) **20 + 11 = 31**
   (B) 20 – 11 = 9
   (C) 20 + 1 + 1 = 22
   (D) 20 – 1 – 1 = 18
   (E) 2 + 0 + 1 + 1 = 4

31 is the greatest number.

2. (C) Wednesday

   The word KANGAROO has 8 letters. Michael will paint the letters on the following days: K on Wednesday, A on Thursday, N on Friday, G on Saturday, A on Sunday, R on Monday, O on Tuesday, and finally the last O on Wednesday.

3. (C)

   The left side of the scale weighs 26 + 12 + 8 = 46 kg, and the right side of the scale is currently loaded with 20 + 17 = 37 kg. To balance the scale, we need to add 46 – 37 = 9 kg to the box on the right side of the scale.

4. (E) six

   Paul got up 2½ hours ago and still has 3½ hours before the train leaves, so he got up 2½ + 3½ = 6 hours before it leaves.

5. (B)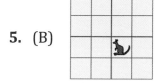

   After the first four moves (one square to the right, one square up, one square to the left, one square down), the toy ended up in exactly the same place as at the beginning. The child then moved it one more square to the right, ending one square to the right from the original location.

6. (A) 3 dollars and 30 cents

   3 scoops of ice cream cost $4.50, so 1 scoop costs $4.50 ÷ 3 = $1.50. 2 cookies cost $3.60, so 1 cookie costs $3.60 ÷ 2 = $1.80. Ala paid for one scoop and one cookie, so she paid $1.50 + $1.80 = $3.30.

7. (B) △

   There are only two gray figures, a triangle and a rectangle. Susan described the one which is not a rectangle. Therefore, she was talking about the gray triangle.

8. (D) 30

   Between 7:45 and 10:45 the clock will strike on the hour at 8:00 (8 times), at 9:00 (9 times), and at 10:00 (10 times). Also, it will strike at half past the hour at 8:30 (1 time), at 9:30 (1 time), and at 10:30 (1 time). The clock will strike 8 + 9 + 10 + 1 + 1 + 1 = 30 times.

SOLUTIONS 2011

9. (C)

   For each figure, count the number of "full" squares and "half-squares."
   The area of 2 half-squares is equal to 1 full square.
   (A) 8 full squares + 4 half-squares = 8 squares + 2 squares = 10 squares
   (B) 8 full squares + 6 half-squares = 8 squares + 3 squares = 11 squares
   (C) 8 full squares + 8 half-squares = 8 squares + 4 squares = **12 squares**
   (D) 8 full squares + 2 half-squares = 8 squares + 1 square = 9 squares
   (E) 8 full squares + 2 half-squares = 8 squares + 1 square = 9 squares
   Figure (C) has the greatest area.

10. (B) 6
    To find the least number of boxes, the larger boxes need to be filled first. 12 eggs fit in a large box, so 5 large boxes will hold 12 × 12 = 60 eggs. The other 6 eggs will fit in 1 small box. The least number of boxes the farmer needs to store 66 eggs is 6.

11. (B) 12
    There are 8 cats, 6 dogs, and 3 fish in the picture. Two students have a dog and a fish, three students have a cat and a dog, so these five students together have 3 cats, 5 dogs, and 2 fish. The remaining 5 cats, 1 dog, and 1 fish, belong one each to 7 students. There are 5 students with multiple pets and 7 students with one pet each, so there are 12 students in the class.

    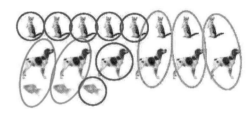

12. (C) 21
    Each cake was divided into 4 × 3 = 12 pieces. There were two identical cakes, so there were 2 × 12 = 24 pieces altogether. Each person got one piece and three pieces were left over, which means that there were 24 – 3 = 21 people at the party.

13. (E) E
    After folding the paper, the right side of the picture will look as shown.

14. (A) 60 cents
    If all 13 coins are 10 cent coins, John has 130 cents. If all 13 coins are 5 cent coins, John has 65 cents, so he cannot have less money than that, such as 60 cents, in his pocket.
    If John has twelve 10¢ coins and one 5¢ coin, the total value is 125 cents.
    If he has ten 10¢ coins and three 5¢ coins, the total value is 115 cents.
    If he has three 10¢ coins and ten 5¢ coins, the total value is 80 cents.
    If he has one 10¢ coins and twelve 5¢ coins, the total value is 70 cents.

15. (D) 5

The possible results on a six-sided die are 1, 2, 3, 4, 5, and 6. If Ari got a result four times as great as Chuck, the only possibility is that Ari got 4 and Chuck got 1. If Darius got a result twice as great as Jack and three times as great as Mark, then it must be that Darius got 6, Jack got 3 and Mark got 2. Tom rolled to only number not mentioned yet, 5.

16. (C) 11

The squirrel will miss either the nut in row 1 column 2, or row 2 column 1 (shaded in pink). It will also not be able to get to the nuts in row 1 column 3, row 1 column 4, row 3 column 1, and row 4 column 1 (shaded in green). Since there are 16 nuts total and it will not be able to get at least 5 of them, the most nuts that the squirrel can get is 16 – 5 = 11. In the first two steps the squirrel can gather 2 nuts and after that the black arrows show one of the two possible paths for gathering 9 more nuts.

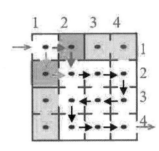

17. (A) 7

If Mrs. Smith had answered exactly half of the questions correctly, her score would stay at 10 points since her wins and losses would cancel each other out. She got 14 points, so she answered more than half of the questions correctly. The incorrect answers were cancelled out by some of the correct answers. In addition to the group of the incorrect and the correct answers which cancel them out, there are 4 more correct answers, because Mrs. Smith ended with 4 more points than she started with. Thus, the incorrect and the correct answers that cancel them out make up 10 – 4 = 6 answers. Half of these answers were incorrect, so Mrs. Smith answered 3 questions incorrectly, which means that she gave 10 – 3 = 7 correct answers.

18. (C) Dasha, Sasha, Pasha, Masha

Take a look at where the girls ended up and work backwards. The order at the end was Masha, Sasha, Dasha, Pasha. There were two changes of places. Since we're working backwards, first we need to look at the last change. Dasha changed places with Pasha, so the order before the initial switch was Masha, Sasha, Pasha, Dasha. In the first change of places, Masha changed places with Dasha, so the original order was Dasha, Sasha, Pasha, Masha.

19. (C) 10

From the diagram, we know that *AF* = 34 and *AD* = 21, so *DF* = 34 – 21 = 13. We also know that *CF* = 23 and since *DF* = 13, *CD* = 23 – 13 = 10. So, the distance between towns *C* and *D* is 10 miles.

© Math Kangaroo in USA, NFP   159   www.mathkangaroo.org

20. (C)

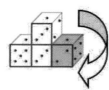

To see what the structure looks like from behind, we rotate it and focus on the colored die shown on the right. It will now be on the left with 1 dot still on the top, the initial front face (2 dots) in the back and the initial back face (5 dots) now in the front. The initial face with 4 dots will now be on the left side of the structure (not visible). The top die after the rotation will show 3 dots as the new front face and 5 dots as the new right face, thus showing structure (C).

21. (E) 12

A card that has a 6 can also be a 9, and a 9 can be a 6. Count all three-digit numbers that have two 6's and one 8, two 9's and one 8, or one 6, one 9, and one 8. These numbers are 668, 686, 866; 899, 989, 998; 689, 698, 869, 896, 968, and 986.

22. (D)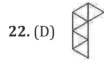

Figure (D) consists of 6 small triangles and the ornament consists of 18 small triangles, so we would need 3 pieces of (D) to build the ornament without any overlap.

There are 6 positions of figure (D) that fit into the ornament, 3 of them can be put into the ornament in two ways which gives 9 configurations shown below.

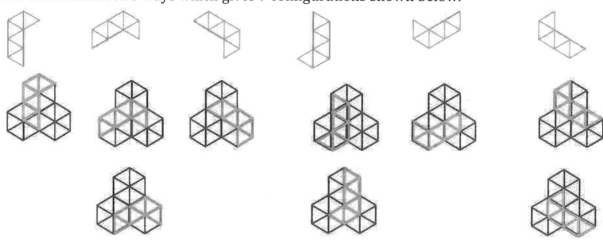

We cannot add two more pieces like that without overlap, so we cannot build the ornament using (D).

The ornament can be built from each of the other shapes as shown below.

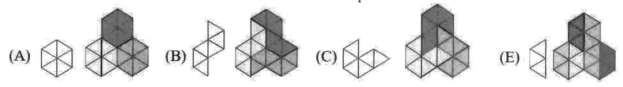

© Math Kangaroo in USA, NFP   160   www.mathkangaroo.org

SOLUTIONS 2011

23. (A) 56
The castle has 8 taller columns made up of 3 cubes each (8 × 3 cubes = 24 cubes) and 16 columns made up of 2 cubes each (16 × 2 cubes = 32 cubes). 24 + 32 = 56 cubes were used to build the castle.

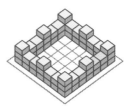

24. (E) 16
8 cannot be placed on the edge with 6 since 8 + 6 = 14 is greater than 13.
8 cannot be placed on the edge with 7 since 8 + 7 = 15 is greater than 13.
If 8 is placed in the white circle opposite 6, then the top edge contains only numbers less than 6, so its sum would not exceed 5 + 4 + 3 = 12, which is less than 13. If 8 is placed in the white circle opposite 7, then the left edge contains only numbers less than 6, so its sum would be less than 13.
Therefore, 8 must be placed in the upper left corner.

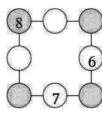

5 cannot be placed on any edge containing 8 since 5 + 8 + a third number is always greater than 13, so the only place for 5 is the lower right corner. After that the other numbers are easy to calculate and place in the proper circles.

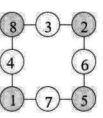

The sum of the numbers in the circles shaded yellow is 8 + 2 + 5 + 1 = 16.

Solutions for Year 2012

1. (B) 8
There are 8 different letters in the word MATHEMATICS: M (twice), A (twice), T (twice), H, E, I, C, and S, so Basil will need 8 colors.

2. (D) ![small square image]

The blue vertical line and the inside horizontal line divide the original square into four corner squares. The upper left square and the lower right square are divided evenly into the white and the gray areas. Each of the other two corner squares show more white than gray area, so in (D) the white area and the gray area are different.

3. (B) 10
4 pins are needed for 3 towels. The next towel will use one pin from a towel that is already on the line and one new pin at the other end, so 5 pins are needed for 4 towels. Any new towel will use one pin which is already on the line at one end and one new pin at the other end, so the father needs one additional pin for each additional towel at this point. Since he needs to hang up 9 – 3 = 6 more towels, he needs 6 more pins, so he needs 4 + 6 = 10 pins total.

© Math Kangaroo in USA, NFP   161   www.mathkangaroo.org

4. (C)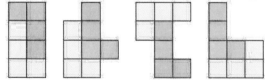
The letter tells us in which column the colored square is, and the number tells us in which row. Among all patterns shown below only (C) has the square B1 colored gray. The other gray squares of (C) are A2, B2, B3, C3, D3, B4 and D4 which, together with B1, are exactly the squares colored by Iljo.

5. (A) 3
One of 13 children playing is the "seeker" and the other 12 children were hiding. 9 children have been found, so 12 − 9 = 3 are still hiding.

6. (E) Mike; he scored 4 points more.
Mike's points are 25 + 35 + 7 = 67. Jake's points are 15 + 45 + 3 = 63.
Mike won; he scored 67 − 63 = 4 points more than Jake.

7. (B) 8
Each row of the rectangular pattern had 4 gray tiles and 4 striped tiles, so there were 4 × 5 = 20 gray tiles. 12 gray tiles still remain, so 20 − 12 = 8 gray tiles have fallen off.

8. (D) on February 24
Ducklings which are 20 days old on March 15 had to hatch 20 days before that day. There are 14 days in March before March 15, so we need to count the last 20 − 14 = 6 days in February. These days are 29th, 28th, 27th, 26th, 25th and 24th, so the ducklings hatched from their eggs on February 24.

9. (E) 4
The construction of all 4 shapes (one piece green, the other gray) is shown below:

10. (B) 6
The difference in cost between 3 balloons and 1 balloon is 12 cents, so 2 balloons cost 12 cents. Therefore, one balloon costs 6 cents.

11. (E) 10
After Grandmother decorated 15 cookies with raisins, there were still 5 plain cookies left. To get the smallest number of cookies decorated with both raisins and nuts, she first has to decorate the 5 plain cookies with nuts. That forces her to decorate 10 more cookies already decorated with raisins, so 10 is the smallest number of cookies decorated both with raisins and nuts.

12. (C) 3

Let's evaluate the expressions as they appear in the sudoku square.

1		3	
4	3		1
3	4	1	
2	1		

In the 2nd row the missing number is 2. Write it in and look to the right.

1		3	
4	3	2	1
3	4	1	
2	1		

Now in the 3rd column the missing number is 4. Write it in and look at the last row.

1		3	
4	3	2	1
3	4	1	
2	1	4	

The missing number in the lower right corner is 3.

1	2	3	4
4	3	2	1
3	4	1	2
2	1	4	3

The complete solution is shown to the left.

13. (D) 25

Nikolay's classmates can be split into groups with 2 girls and 1 boy in each group. Each group consists of 3 of Nikolay's classmates, so the number of all children in this class (excluding Nikolay) is a multiple of 3. Subtract 1 from the possible answers and check which result is a multiple of 3. 30 – 1 = 29, 20 – 1 = 19, 24 – 1 = 23 and 29 – 1 = 28 are not multiples of 3. Only 25 – 1 = 24 is a multiple of 3 (24 = 8 × 3). 25 can be the number of children in the class.

14. (B) 5

3 kittens have 3 × 4 = 12 legs, 4 ducklings have 2 × 4 = 8 legs, and 2 baby geese have 2 × 2 = 4 legs. Together these animals have 12 + 8 + 4 = 24 legs, so the lambs have 44 – 24 = 20 legs. Since each lamb has 4 legs, there were 20 ÷ 4 = 5 lambs in the animal school.

15. (D)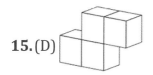

 At least one face of each light gray cube is shown. The same is true about each dark gray cube. One black cube is hidden but it is connected to the 3 black cubes shown. The only space for it is under the white cube which is adjacent to the black cube (see the top face of the prism). There are only two spaces available for the two hidden white cubes, one under the other white cube visible on the top face of the prism and the other next to it under the dark gray corner cube at the top and back of the prism. The shape of the white piece is shown in (D).

16. (C) 57

There were 6 five-branched candlesticks and 15 – 6 = 9 three-branched candlesticks, so 6 × 5 + 9 × 3 = 30 + 27 = 57 candles had to be bought for all the candlesticks.

17. (D) 12

The grasshopper can jump up 7 times reaching the 21st step. At this point she can jump down landing on the 17th step. Now, she can jump up reaching the 20th step. The next jump must be down to the 16th step and from there she will make 2 jumps up to reach the 22nd step.
3 + 3 + 3 + 3 + 3 + 3 + 3 – 4 + 3 – 4 + 3 + 3 = 22 and 10 jumps up + 2 jumps down is 12, which is the smallest number of jumps to take a rest on the 22nd step.
The grasshopper can change the order of up and down jumps as long as she does not go below the ground level.

18. (C) 4

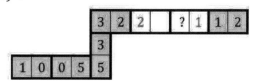

The number of dots on each tile is shown here by numbers. Adjacent parts of different tiles have the same numbers, so the left missing tile has 2 dots on its left side and the right missing tile has 1 dot on its right side (see both in red).
The sum of the dots already accounted for is 25 = 1 + 0 + 0 + 5 + 5 + 3 + 3 + 2 + 2 + 1 + 1 + 2.
3 – 25 = 8, and since the two missing tiles have the same number touching, the ? has 4 dots.

19. (D) 1173

To create the largest sum, the numbers will have the largest two digits (6 and 5) in the hundreds place, the next largest two digits (4 and 3) in the tens place, and the smallest two digits (2 and 1) in the ones place. There are a few ways to do this (for example 632 and 541), all of which give the same sum of 1173.

20. (B) 4

If Laura stands to the right of Kate, then Iggy will stand to the right of Laura. Val has two ways to be in the picture, at either end of the group.
If Laura stands to the left of Kate, then Iggy will stand to the left of Laura. Again, Val has two ways to be in the picture, at either end of the group.
This group can pose for the picture in 4 different ways: VKLI, KLIV, VILK, or ILKV.

21. (E)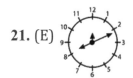

12:55:30 is almost one o'clock, so the medium hand is the hour hand. At 55 minutes past the hour the minute hand will be pointing to the 11, so the longest hand is the minute hand. The shortest hand shows the seconds.
At 8:11:00 the hand showing seconds points up directly toward 12, so the above options (A), (B) and (D) are eliminated right away. At 8:11:00 the minute hand must be a little past the 10 minutes mark represented by 2 on the clock. This also eliminates (C). (E) shows the hour just a bit past eight o'clock, so that is the clock that shows 8:11:00.

22. (D) 7

Let's reverse the process. Since Michael multiplied by 4 in the last operation to get the final answer of 2012, we need to divide 2012 by 4, which results in 503. Since he added 3 in the previous step, we need to subtract 3 from 503, which equals 500. Before that, he multiplied his previous result by 10, so we will divide 500 by 10, which gives us 50. He added 1, so we will subtract 1 from 50, which results in 49. The number which Michael chose at the beginning needs to result in 49 when multiplied by itself. That number is 7, because 7 × 7 = 49.

23. (E) 12 mm

From the 84 × 192 rectangle we can cut off two 84 × 84 squares since 84 + 84 = 168 is less than 192. What remains is a 84 × 24 rectangle since 192 − 168 = 24.
From the 84 × 24 rectangle we can cut off three 24 × 24 squares since 24 + 24 + 24 = 72 is less than 84. What remains is a 12 × 24 rectangle since 84 − 72 = 12. The 12 × 24 rectangle can be split into two 12 × 12 squares, and now there are no more non-squares to cut, so 12 mm is the length of the side of the smallest square.

In the picture above, the two 84 × 84 squares are shown in blue and orange. The 84 × 24 rectangle is shown with the black perimeter. The three 24 × 24 squares are shown in red, green and purple. The two 12 × 12 squares are shown in yellow and white.

24. (C) 10

Any time there are 3 or more ties, we can increase the number of losses by replacing 3 tied games by 1 winning game and 2 losing games. The number of games is still the same since 1 + 2 = 3 and the number of points is also the same since 3 + 0 + 0 = 1 + 1 + 1. For the greatest number of losses, the number of ties must be either 0, 1, or 2.
If there were no ties, then 80 points would be equal to 3 points × the number of wins. This cannot happen since 80 is not a multiple of 3.
If there were only one tie, then 80 − 1 = 79 points would be equal to 3 points × the number of wins. It cannot happen since 79 is not a multiple of 3.
The only option left is 2 ties, so the number of wins is 26 since 3 × 26 + 2 = 80. In this case, the number of lost games is 38 − (26 + 2) = 10, so 10 is the greatest possible number of games that the team lost.

Solutions for Year 2013

1. (D)

   Figure (A) shows 3 black kangaroos and 3 white kangaroos.
   Figure (B) shows 4 black kangaroos and 4 white kangaroos.
   Figure (C) shows 4 black kangaroos and 4 white kangaroos.
   Figure (D) shows 5 black kangaroos and only 4 white kangaroos.
   Figure (E) shows 5 black kangaroos and 5 white kangaroos.

   Figure (D) is the only figure with more black kangaroos than white kangaroos.

2. (D) **7**

   The sum of the two digits needs to end in 4. The options are 2 + 2 = 4 and 7 + 7 = 14. However, using 2 does not makes the calculation correct: 42 + 52 = 94, not 104. 47 + 57 = 104, so 7 is the digit under the stickers.

3. (E) ⬤⬤⬤⬤

   The pattern is such that after a certain number of black circles, there is the same number of gray circles, so the last 4 circles should all be shaded gray.

4. (B) 10
   There are ten triangles in total, eight small ones numbered 1 to 8, and two big ones numbered 9 (red) and 10 (purple).

   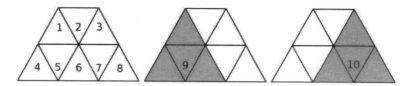

5. (C) 16
   USA won a total of 46 + 29 + 29 = 104 medals, while China won a total of 38 + 27 + 23 = 88 medals, which gives a difference of 104 − 88 = 16 medals.

6. (D) 5
   36 is divisible by 2, 3, 4 and 6 from the numbers listed, so Daniel can divide all the candy evenly among those numbers of friends. 36 is not divisible by 5. If Daniel tried to divide the candy equally among five friends, he would have to give each friend seven pieces, and have one piece left over.

SOLUTIONS 2013

7. (B) 30
Since each sandwich needs two slices of bread, Vero's mom can make 12 sandwiches from one package of bread. She can make 12 + 12 + 6 = 30 sandwiches from two and a half packages.

8. (E) Eddie
One of the digits of 325, the digit "2," is not odd, so Eddie was wrong.

9. (B)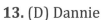

Extend the broken mirror to its original shape to see the missing piece shown in light blue. Rotate the missing piece around to match it with the piece (B).

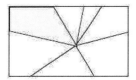 

10. (D) 23 cm
Since three lies add 3 × 6 cm = 18 cm to the length of his nose, and two true statements shorten it by 2 × 2 cm = 4 cm, the final length of Pinocchio's nose is 9 cm + 18 cm – 4 cm = 23 cm.

11. (D) 5
Pedro cannot buy just four boxes of oranges since buying four boxes of the largest size would give him only 40 oranges. However, if Pedro buys three boxes of 10 and two boxes of 9 oranges, then he will have exactly 3 × 10 + 2 × 9 = 48 oranges in five boxes.

12. (A)
Ann's turns have been marked in the figure to the right. After all of the shown turns, she is walking towards the train.

13. (D) Dannie
Betty and Andy were both born in the same month, which happens to be in May. Andy and Cathie were born on the same day, so it must have been on the 12th. Andy was born in May, so Cathie was born April 12th. This means Dannie was born on February 20th. The person whose birthday comes first is the oldest, so Dannie is the oldest.

14. (E) 5
Each of the 30 children at Adventure Park participated either in the "moving bridge" contest, went down the zip-line, or took part in both events. 20 children went down the zip-line, so 30 – 20 = 10 did not go down the zip-line but definitely participated in the "moving bridge" contest. There are 15 – 10 = 5 other children who participated in the "moving bridge" contest, so these 5 children took part in both events

15. (B)

Piece (B) fits with the original piece as shown to the right.

16. (B) 3
22 is divisible by 2, 24 is divisible by 4, and 25 divisible by 5. There are 3 numbers greater than 21 and smaller than 30 that have this property.

17. (D) 16
After the first step, we get the smaller red triangle and 3 others of the same size together covering the original triangle. Repeating the same steps with the red triangle we get an even smaller blue triangle and 3 other triangles of that size; together they cover the red triangle. We can do this to all the other triangles of the size of the red one, so the number of all the triangles of the size of the smallest resulting triangle (the blue one) that fit in the original triangle is 4 × 4 = 16.

18. (D) 102
Each number from 2013 to 2110 has at least one 0 as a digit, so the product of the digits is always 0 and is smaller than the sum of digits (the sum is greater than 2). 2110 is followed by 2111, 2112, 2113, 2114, and 2115. 2115 is the first year that has the product of digits greater than the sum of digits since 2 × 1 × 1 × 5 = 10 is greater than 2 + 1 + 1 + 5 = 9.
2115 − 2013 = 102, so 102 years will pass before the product of the digits in the notation of the year is greater than the sum of these digits.

19. (B) (31 − 7 × 3) × 24 × 60
There are 31 days in December and she slept 3 weeks or 7 days × 3, so the number of days she was awake is (31 − 7 × 3) days. Each day has 24 hours and each hour has 60 minutes, so she was awake for (31 − 7 × 3) × 24 × 60 minutes.

20. (C) 5
Among Basil's seven tiles there are the following squares: 3 single dot, 3 two-dot, 3 three-dot, 3 four-dot, 1 five-dot, and 1 six-dot. The tiles can only be arranged side by side if the numbers of dots on neighboring squares can be paired up. There are 4 such pairs of connecting tiles, one pair for each single dot, two-dot, three-dot, and four-dot squares, so the largest number of tiles Basil can arrange is 5. One of the possible arrangements is shown below.

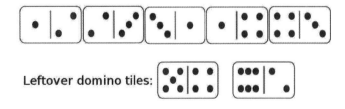

21. (E) Such division is not possible.
The price for all ten glass bells is 1 + 2 + 3 + 4 + 5 + 6 + 7 + 8 + 9 + 10 = 55 dollars. 55 is not a multiple of 3, so Cristi cannot divide all the glass bells into three packages so that each of the packages has the same price.

22. (B) 67
There are 9 squares along the width of 36 inches, so one square has the size of 4 × 4 inches. There are 15 squares along the length of 60 inches since 15 × 4 = 60. There are 15 vertical stripes (columns) in the fully unrolled rug. The first column has 4 moons, the second 5, the third 4, and it repeats itself so that 8 columns have 4 moons, and 7 columns have 5 moons. The total number of moons is 8 × 4 + 7 × 5 = 67.

23. (B) 3
Baby Roo can write one-digit, two-digit, three-digit, and four-digit numbers using only 0 and 1. He must have used at least 3 of those numbers, by adding 1 three times, to get the ones digit of 2013. The only two possible sums of numbers Baby Roo could have written down are either 1 + 1001 + 1011 or 11 + 1001 + 1001.

24. (B) 4
The gray piece consists of 9 small squares but itself is not a square. Two gray pieces consist of 18 small squares and three gray pieces consist of 27 small squares. No square can be made of either 18 or 27 identical small squares, because the number of squares is also the lengths of the sides multiplies by each other, and these number can't be made by multiplying a number by itself. Four gray pieces consist of 36 small squares and the square of the size 6 × 6 also consists of 36 small squares, so there is a chance that using four gray pieces we can build a square. It can be done as shown here, so at least 4 gray pieces are needed to make a completely full gray square.

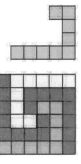

Solutions for Year 2014

1. (D)

The large figure with the star has 9 spikes. In the picture below the number of spikes in each figure was counted. The only figure with 9 spikes is (D), so it is the only figure that could be the central part of the larger figure.

SOLUTIONS 2014

2. **(D) between the 1 and the 4**
   To make the smallest number possible, the smaller digits need to come before larger digits. The digits 2, 0, and 1 are smaller than 3, so 3 should be placed after those digits. 4 is larger than 3, so 3 should be placed before 4. So, Jackie should place the digit 3 between the 1 and the 4 to make the number 20,134, which is the smallest number she can make.

3. **(A) 1, 4**

   House 5 has a brown triangle roof different from any other house, House 2 is the only house made with a smaller yellow square, and House 3 is the only house with a small blue rectangle. Each of these three houses is made with different pieces than any other house.
   Only houses 1 and 4 are made using exactly the same triangular and rectangular pieces.

4. **(D) 200**
   Yesterday Koko was awake for 24 − 20 = 4 hours. Since he eats 50 gram of leaves per hour, during the 4 hours he was awake he ate 50 × 4 = 200 grams of leaves.

5. **(A)**

   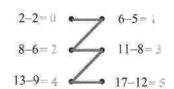

   The results of all the subtractions performed by Maria are marked in green. The red line connects subtraction results from 0 to 5, in increasing order. The figure that Maria draws is seen in (A).

6. **(E) Lucy**
   In the picture below, the names of the children are lined up according to the number of sandcastles they built. Of all 5 children, Lucy built the most sandcastles.

   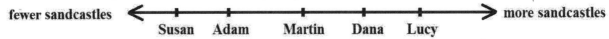

7. **(E) 8**
   The picture below shows all the steps needed to find the number Monica wrote in the gray cell. The number written in the gray cell is 8.

   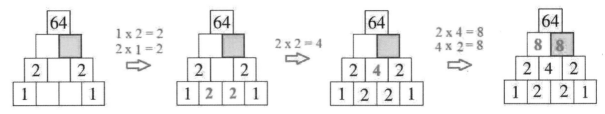

SOLUTIONS 2014

8. (C)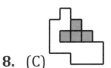
Each of the four pieces is made of 4 small squares. The longest piece is 4 small squares long. The white shape has two columns and one row (shown below in red) of exactly that length. In the first case, there is not enough area to the left of the red piece to place there any piece made of four small squares. In the second case, the area to right has 5 small squares, more than needed for one piece but not enough for two pieces. The third case looks more promising. The T-shaped piece fits exactly in the area above the red row and the area below has 4 + 4 small squares. However, the two pieces cannot fit together in the bottom part of the white shape.

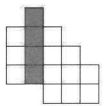 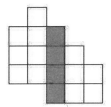 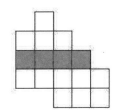

There are two other options to place the longest of the four pieces, shown in red, in the second row from the bottom. The first option shows only 3 small squares below the red piece, so no other piece can fit there. The second option shows 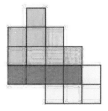 below the long red piece.

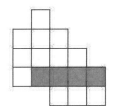 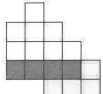

There is only one way to place the other two pieces as shown to the right.
The T-shaped piece is placed in the position shown in (C).

9. (E)
From the other side of the window, the figure will look like its mirror image. The picture to the right shows the original painting and its mirror image, which is the flower seen in (E).

mirror image

© Math Kangaroo in USA, NFP 171 www.mathkangaroo.org

10. (E) 48

In the end there were 6 pieces of candy left in the bowl, so Clara took 6 pieces out, which is half of the 12 pieces that were there before. Tom took 12 pieces of candy out of the bowl, which is half of the 24 pieces that were there beforehand. Sally took 24 pieces out of the bowl, half of the 48 pieces that were there at the beginning.
This process is also shown in the diagram below.

11. (B)

The original picture with one tile missing has 3 tiles exactly half dark gray and half light gray. There are 2 dark gray tiles matched by 2 light gray tiles and 1 light gray tile left over. That left over light gray tile must be matched by one full dark gray tile which is the one that must be added to the picture.

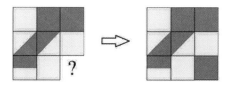

12. (D) 90

When missing the target both times, Paula gets 0 points. When hitting the target only once, she can get 30, 50, or 70 points. If she hits the same area twice, Paula can get 60, 100, or 140 points. If she hits two different areas, Paula can get 80, 100, or 120 points. From the list of the sums given in the answers, only 90 points cannot happen.

13. (B) 6

We can see 4 yellow tokens, 4 green tokens, and 5 red tokens (one is in the middle). Altogether 4 + 4 + 5 = 13 tokens are visible. Mary also has 5 tokens that were not used, so at the beginning she had 13 + 5 = 18 tokens. Since she had equal numbers of tokens in each of the 3 colors, she had 18 ÷ 3 = 6 tokens of each color, so Mary had 6 yellow tokens at the beginning.

14. (B) 7

Peter Rabbit ate 30 carrots in a week by eating either 9, 4 or 0 carrots per day. If he ate only 4 carrots per day, he could not eat 30 carrots during the week since 30 is not a multiple of 4. If he ate only 9 carrots per day, he could not eat 30 carrots during the week since 30 is not a multiple of 9. Therefore, there was one day when he ate 4 carrots and another day when he ate 9 carrots. During the other five days he ate the other 30 − 4 − 9 = 17 carrots. 17 is not a multiple of 4 and 17 is not a multiple of 9, so there were two other days when Peter Rabbit ate 4 + 9 = 13 carrots. He had to eat 17 − 13 = 4 more carrots on another day. Thus, there were two days when Peter Rabbit ate 9 carrots per day and no cabbages, and three days when he ate 4 carrots and one cabbage each day. There were also two days when he ate 2 cabbages each day and no carrots. So, during the week Peter Rabbit ate 2 × 0 + 3 × 1 + 2 × 2 = 7 cabbages.

SOLUTIONS 2014

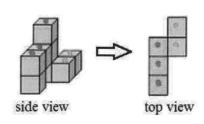
side view   top view

15. (C)
From directly above, only the sides marked with red and green dots will be visible. So, when looking directly from above the solid will look like shape (C).

16. (B) 181
We can color squares in the 4 non-overlapping rows. Each of these rows has exactly 8 squares, so there are 4 × 8 = 32 colored squares. Each of them has 5 dots, so there are 32 × 5 = 160 dots inside the colored squares. There are 3 rows of dots between the colored rows, each with 7 dots, so there are 3 × 7 = 21 such dots. The total number dots in the picture is 160 + 21 = 181.

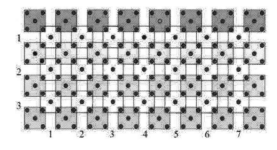

17. (B) 30
A kangyear has 20 kangmonths, so there are 5 kangmonths in one quarter of a kangyear since 4 × 5 = 20. Each kangmonth has 6 kangweeks, so there are 5 × 6 = 30 kangweeks in one quarter of a kangyear.

18. (C) 4 is the only possible number.
A group of 7 girls standing in a circle violates the rule that no three girls are standing next to each other. Same is true for a group of 6 girls and one boy. To avoid another violation of the above rule, a group of 5 girls must be split by two boys into two smaller groups. One of these smaller groups must have at least 3 girls, again violating the rule that no three girls are standing next to each other.
2 or more boys standing as part of the circle divide the circle into as many parts as the number of boys. No two boys are standing next to each other, so there is at least one girl in each part of the divided circle. Thus, the whole group of children must have at least as many girls as boys. In the case of 7 children, the number of boys is at most 3, so the number of girls is at least 7 – 3 = 4. We already know that more than 4 girls would violate the rule that no three girls are standing next to each other.
In conclusion, the only possible options is that in the group of seven children exactly 4 are girls, so 4 is the only possible number.
The picture to the right shows 4 girls and 3 boys standing in the circle according to both rules.

© Math Kangaroo in USA, NFP   173   www.mathkangaroo.org

19. (B) 3

Only two cards are in their right places, so 6 cards need to be moved. In one move we can switch only two cards, so at least 3 moves are needed to switch 6 cards. Indeed, it can be done in 3 moves as shown below.

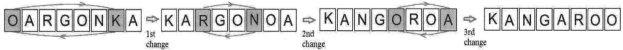

20. (C) 26

All 6 steps are shown in the picture below. The numbers in red show the total number of diamonds in the figures above them. There are 3 diamonds used in step 1. In step 2, the bottom row with 3 diamonds was added. For step 3, the bottom row with 4 diamonds was added. The pattern continues. Step 6 will consist of 28 diamonds with 2 white diamonds and 26 black diamonds.

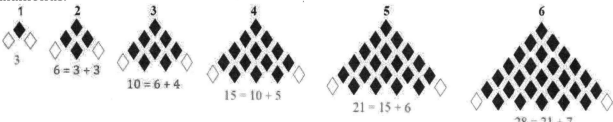

21. (A) the bear and the horse

At first, Hamish bought some toys for 150 − 20 = 130 Kangcoins. For three toys he would pay at least 40 + 48 + 52 = 140 Kangcoins, so he bought exactly two toys. Only 73 + 57 = 130 Kangcoins, so at first Hamish bought the bear and the duck. After the exchange he got back 5 Kangcoins, so he exchanged one of his toys for another toy that was 5 Kangcoins cheaper. Only the horse is 5 Kangcoins cheaper than the duck, so Hamish left the store with the bear and the horse.

22. (D) 5

When we add two two-digit numbers with all different digits that are 6 or less, the sum is at most 117, since 64 + 53 = 63 + 54 = 117. So, the hundreds digit of the sum is 1 and its tens digit has to be 0, since 1 cannot be used twice. The two addends (numbers being added) can use only the remaining digits 2, 3, 4, 5, and 6. Using only 2, 3, 4, and 5, would make the sum at most 95 since 53 + 42 = 95 or 52 + 43 = 95. This sum would never be a 3-digit number. Thus, 6 has to be a digit of one of the addends. 6 cannot be the ones digit of either addend since 6 + 2 = 8, 6 + 3 = 9, 6 + 4 = 10, and 6 + 5 = 11. The ones digits of these results are 8, 9, 0, and 1. However, 0 and 1 are already used, and 8 and 9 are not on the list of digits. For the same reason 5 cannot be the ones digit of either addend. 5 and 6 cannot be the tens digits at the same time since the sum would be greater than 110 but we already know that the tens digit of the sum is 0. Thus 5 has to be placed in the shaded square. Ignoring the order of the addends, the possible solutions are displayed below.

23. (D) 21

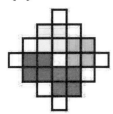

There are four 2 × 2 non-overlapping squares shown in different colors in the figure on the left. Each of them must have at least one unshaded unit square, so the figure must have at least 4 unshaded small squares. The figure has 25 small squares, so we cannot shade more than 25 − 4 = 21 of them. As shown to the right it is possible to shade exactly 21 small squares.

24. (D) 8

The shaded cell has 4 neighboring cells. In these cells we can put only numbers greater than 4. The sum of numbers in these 4 neighboring cells is never 13 since it is greater than 4 × 4 = 16. Therefore, neither 5 nor 6 can be written in the shaded cell. Only 7, 8 or 9 can be written in that cell.

Each of the four middle edge cells has three neighboring cells. In each case, the sum of numbers in the three neighboring cells depends on the number in the shaded cell.

When 7 is written in the shaded cell, the sums of the middle and corner cells that would be the neighbors for 5 or 6 are:
1 + 2 + 7 = 10,   2 + 3 + 7 = 12,   3 + 4 + 7 = 14   and   4 + 1 + 7 = 12.
None of the results is 13, so 7 cannot be written in the shaded cell.

If we replace 7 by 9, then the corresponding sums are
1 + 2 + 9 = 12,   2 + 3 + 9 = 14,   3 + 4 + 9 = 16   and   4 + 1 + 9 = 14.
Again, none of the results is 13, so 9 cannot be written in the shaded cell.

8 is the only option left and the corresponding sums are
1 + 2 + 8 = 11,   2 + 3 + 8 = 13,   3 + 4 + 8 = 15   and   4 + 1 + 8 = 13.
Two of the sums are equal to 13, so 5 and 6 (in any order) can be written in the cells between 2 and 3 and between 4 and 1. In either case, Nick wrote 8 in the shaded cell.

Solutions for Year 2015

1. (E) 15

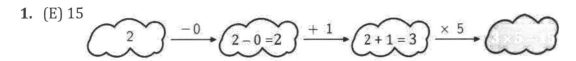

2. (A) A

The longest connected strip Eric made was the one with the smallest number of holes between the two screws. The number of holes is shown next to each long strip. Strip A is the longest.

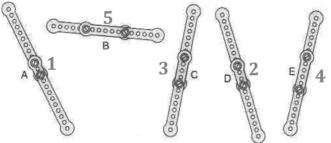

3. (E) 6
   The number hidden behind the triangle is 3 since 3 + 4 = 7.
   The number hidden behind the square is 6 since 6 + 3 = 9.

   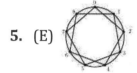

4. (C) (1000 − 1) ÷ 9
   Evaluate each expression.
   (A): (1000 − 100) ÷ 10 = 900 ÷ 10 = 90
   (B): (1000 − 10) ÷ 9 = 990 ÷ 9 = 110
   (C): (1000 − 1) ÷ 9 = 999 ÷ 9 = 111
   (D): (1000 − 100) ÷ 9 = 900 ÷ 9 = 100
   (E): (1000 − 10) ÷ 10 = 900 ÷ 10 = 90
   111 is the largest number, so (1000 − 1) ÷ 9 has the greatest value.

5. (E)

   The dots are connected in the following order 1 → 3 → 5 → 7 → 9 → 2 → 4 → 6 → 8.
   All dots are connected in the pattern which is shown in figure (E).

6. (E) 8
   Since the product of the two digits is 15, the only possibility for these two digits is 3 and 5. The sum of the digits is 3 + 5 = 8. It does not matter whether the number is 35 or 53.

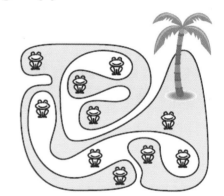

7. (B) 6
   The palm tree is growing on the island. Shade the land area connected to the palm tree and count the frogs in the shaded area. There are 6 frogs sitting on the island.

8. (D) Saturday
   It will be Thursday again in 7, 14, 21, and 28 days. In 29 days it will be Friday, and 30 days from this Thursday it will be Saturday.

9. (A)
   KANGAROO is spelled clockwise (looking from above) on the umbrella, so the only legitimate three-letter pieces of the umbrella are: KAN, ANG, NGA, GAR, ARO, ROO, OOK and OKA. The only match is NGA in the figure (A).

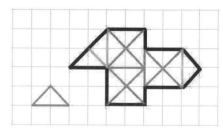

10. (D) 15
    15 identical non-overlapping triangles of the given size are covering the shape.

SOLUTIONS 2015

11. (B) 3

After giving 2 apples to Yuri, Luis had 7 – 2 = 5 apples left. Luis also had 2 bananas, so Yuri gave Luis 3 bananas, since 2 + 3 = 5, and now Luis has 5 of each fruit.

12. (B) 4

Each grandchild received 4 pieces of candy and Grandma had 2 pieces left. With 2 more pieces she would be able to give each grandchild 1 more piece of candy. 2 + 2 = 4, so with 4 extra pieces of candy, giving 1 more piece to each grandchild means that Grandma has 4 grandchildren. Check: 4 × 4 + 2 = 18 and 4 × 5 – 2 = 18.

13. (C) 4

Not counting Tom, there are 9 skaters. The number of skaters who came after Tom is 3 more than the number of skaters who came before him. Removing those 3 extra skaters, leaves 9 – 3 = 6 skaters with equal numbers before and after Tom. Thus, 3 skaters finished the competition before Tom. Tom ended up in 4th place.

14. (B) 4

Since both the ship and the airplane have to be next to the car, there are two possible arrangements for these 3 toys: either Ship–Car–Airplane or Airplane–Car–Ship. The ball can be placed either to the left or to the right of these two arrangements, giving us 4 possible ways in which the toys can be placed: Ball–Ship–Car–Airplane, Ball–Airplane–Car–Ship, Ship–Car–Airplane–Ball, or Airplane–Car–Ship–Ball.

15. (D) D

Draw over the picture following the instructions for making turns. Notice that Peter never passes point D.

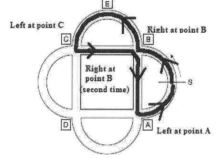

16. (C) 6

There are 5 ladybugs: one ladybug with 2 spots, two ladybugs with 3 spots, one ladybug with 5 spots, and one ladybug with 6 spots. Since each ladybug sent one text to a ladybug whose number of spots differs exactly by 1, it means that the texts were exchanged between 2-spotted and 3-spotted ladybugs, and separately between 5-spotted and 6-spotted ladybugs. The ladybug with 2 spots sent one text to each of the two ladybugs with 3 spots (2 texts) and received a text back from both (2 texts). The ladybug with 5 spots sent a text to the ladybug with 6 spots and received a text back (2 texts). Altogether 6 text greetings were sent.

17. (A)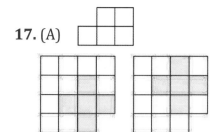

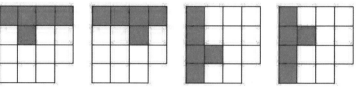

Piece (E) fits nicely in the cut off corner but no other corner of the figure can be covered by (E).

The upper right corner can be covered by piece (D) only when the longer side of (D) is horizontal. The lower left corner can be covered by piece (D) only when the longer side of (D) is vertical. One piece (D) in a horizontal position and another one in a vertical position always overlap, so the figure cannot be covered by non-overlapping pieces of (D).

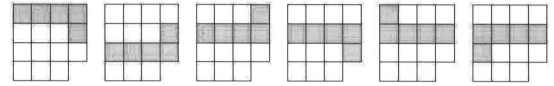

The middle square of the shorter column can be covered by piece (C) when the longer side of it is placed horizontally. By symmetry, the middle square of the shorter row can be covered by piece (C) when the longer side of it is placed vertically. One piece (C) in a horizontal position and another one in a vertical position always overlap, so the figure cannot be covered by non-overlapping pieces of (C).

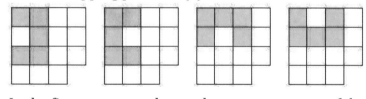

The upper left corner of the figure can be covered in four different ways by piece (B). The same is true for the lower left and the upper right corners.

In the first two cases shown above any coverage of the lower left corner overlaps each coverage of the upper left corner. In the next two cases any coverage of the upper right corner overlaps each coverage of the upper left corner, so the figure cannot be covered by non-overlapping pieces of (B).

As shown to the right, the figure can be divided into three identical pieces that look like (A). We can use symmetry to get another arrangement.

18. (C) 13

A cube has 4 edges in each of the three directions, so it has 12 edges altogether. Along each edge there is exactly one small white cube, so there are 12 small white cubes along all edges. Another small cube is at the center of the big cube. It is white since it is adjacent to the gray middle cubes of the six faces. Thus, Jack used 12 + 1 = 13 white cubes.

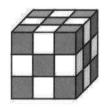

SOLUTIONS 2015

19. (A) 4
   Let's call the small pitcher S, the medium pitcher M, and the large pitcher L. There are two ways to fill in the barrel with water: the first option is to use 6S + 3M + 1L, the second option is to use 2S + 1M + 3L. A clever way to approach this problem is to notice the proportion of small and medium pitchers for the two options of filling the barrel. In the first option it is 6S + 3M and in the second option it is 2S + 1M. If we triple the number of small and medium pitchers in the second option, 3 × (2S + 1M), we get a quantity equivalent to the first option (6S + 3M). Let's consider what it would take to fill 3 barrels. Using the second option 3 × (2S + 1M + 3L), it would take 6S + 3M + 9L to fill 3 barrels, which we can express as (6S + 3M + 1L) + 8L. From the first option, 6S + 3M + 1L is equivalent to 1 barrel, so 8 large pitchers are equivalent to 2 barrels. To fill a barrel, we would need 4 large pitchers.

20. (D) 5 or 7
   After picking one out of the 5 given numbers (2, 3, 5, 6, and 7) to be placed in the center square, the remaining 4 numbers must form pairs where the sum of the smallest and the largest is equal to the sum of the other two. We can eliminate both even numbers (2 or 6) from being placed in the center square, since the remaining 4 numbers would consist of one even and three odd numbers, which could not form equal sums. Placing 3 in the center square would not work, since 2 + 7 ≠ 5 + 6. Placing 5 in the center square makes the sums 2 + 7 and 3 + 6 equal. Placing 7 in the center square would also work, since 2 + 6 = 3 + 5.
   In conclusion, either 5 or 7 can be placed in the center square of the cross.

21. (E) 15
   For the product to equal to 0, one of its factors must be a 0. This is one of Peter's three numbers.
   The product of Ann's, George's, and Peter's numbers (without 0) is the product of all digits from 1 to 9. That product is 1 × 2 × 3 × 4 × 5 × 6 × 7 × 8 × 9 and, with no references to balls, can be written as 2 × 4 × 7 × 3 × 5 × 6 × 8 × 9 = 56 × 90 × 72. 90 is the product of Ann's numbers and 72 is the product of George's numbers, so 56 must be the product of Peter's two nonzero numbers less than 10. The only option for Peter is 56 = 7 × 8, so the sum of his three numbers is 0 + 7 + 8 = 15. For Ann and George there are two options 90 = 9 × 5 × 2 and 72 = 6 × 4 × 3 × 1 or 90 = 6 × 5 × 3 and 72 = 9 × 4 × 2 × 1.

SOLUTIONS 2015

22. (C)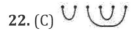

Assign numbers from 1 to 6 (from left to right) to the ends of the ropes. For the original three ropes the ends are 1 & 4, 2 & 6 and 3 & 5.

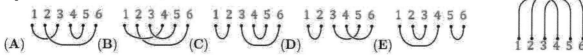

Set (A) will create a short closed loop with the ends 2 & 6. Set (B) will create three short closed loops with the ends 1 & 4, 2 & 6 and 3 & 5. Set (D) will create a short closed loop with the ends 3 & 5. Set (E) will create a short closed loop with the ends 1 & 4. Only set (C) creates a loop that will make one complete rope.

(A)   (B)   (C)   (D)   (E)

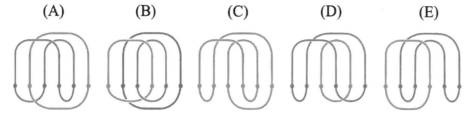

23. (D) 8

Keep the top square as it is and rotate the other two squares. There are 4 different configurations of the middle square obtained by its rotations shown in a row below. Due to the symmetry, there are only 2 different configurations of the bottom square obtained by its rotations shown in a column below.

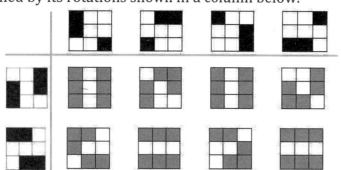

There are 2 × 4 = 8 squares when the two column configurations are put on the top of the four row configurations. Below the same matches are displayed when the top square is added.

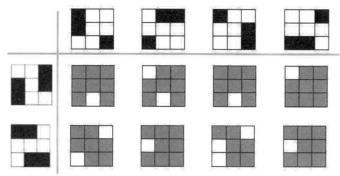

The maximum number of black squares is 8.

SOLUTIONS 2015

24. (C) Charlie
    Anna made 6 times as many cookies in two days as she made on Friday, since among the numbers 24, 25, 26, 27 and 28, only 24 is a multiple of 6, so Anna made 4 cookies on Friday. Berta made 5 times as many cookies in two days as she made on Friday, since only 25 is a multiple of 5, so Berta made 5 cookies on Friday. Elisa made 4 times as many cookies in two days as she made on Friday, since only 24 and 28 are multiples of 4, so Elisa made 7 cookies on Friday. David made 3 times as many cookies in two days as he made on Friday, since only 24 and 27 are multiples of 3, so David made 9 cookies on Friday.
    Finally, Charlie made twice as many cookies in two days as he made on Friday, so Charlie made 13 cookies on Friday. Charlie baked the most cookies on Friday.

Solutions for Year 2016

1. (E) Ernst
   The totals are:
   Amy: 6 + 1 = 7
   Bert: 3 + 3 = 6
   Carl: 2 + 3 = 5
   Doris: 4 + 4 = 8
   Ernst: 5 + 4 = 9
   The largest total is 9. This is what Ernst rolled.

2. (E) 5
   8 weeks = 7 weeks + 2 days + 5 days, so Little Kanga will be 8 weeks old in 5 days.

3. (A) 24
   17 + 3 = 20, 20 − 16 = 4, and 20 + 4 = 24.

4. (A)
   An experiment shows a mirror image of ⊢ as ⊣, so objects are reversed left-to-right but not top-to-bottom. It is like looking at a flat picture on the front of a transparent paper from the **back** of the paper.

5. (D) ⬈
   71 and 72 are seats between 61 and 80. It is indicated by ⬈.

   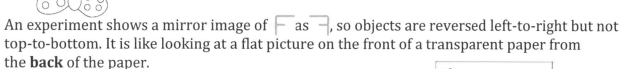

6. (B) 3
   There are 6 people involved, Anna + 5 friends. 6 × half of an apple = 3 apples.

7. (A) a triangle

Extend the shaded region to a full rectangle to see a triangle behind the curtain.

8. (C) There are twice as many circles as triangles.

There are 4 circles, 2 squares, and 2 triangles. 4 = 2 × 2, so there are twice as many circles as triangles which makes the statement (C) true. The other statements are false.

9. (B) 2025
If the sum of digits of any year is 9, then the year must be a multiple of 9.
2016 is a multiple of 9, so we look at the next multiple of 9, 2016 + 9 = 2025, and check that 2 + 0 + 2 + 5 = 9.
(Be aware that we were just lucky that the very next multiple worked. If the starting year were for example 2070, then the next multiple of 9 would be 2070 + 9 = 2079, with the sum of the digits equal to 18, not 9. The first year after 2070 with the sum of its digits equal to 9 will be 2106.)

10. (B) 4

The mouse can't go back through any gate, so it must pick (in the first move) one of the purple gates and then one of the green gates. 2 × 2 = 4, so there are 4 different path choices (yellow, black, blue, and red).

11. (C) 11 and 4
On each of two cards Zoe has a sum of two numbers. By adding them, we are adding all four numbers, so the total of the two sums is 32. These two sums are equal, so the sum on each card is 32 ÷ 2 = 16. The numbers we cannot see are 16 − 5 = 11 and 16 − 12 = 4.

12. (B)

Blue is the only path which connects adjacent sides of the middle tile. This means that only tiles (B) or (C) can possibly work. Notice that in the picture tile (B) fits directly in the middle, whereas even after rotating of tile (C) so that the blue path in the upper left corner, the red and yellow paths are switched.

13. (D) 3
The circle doesn't cover any vertex of the square or the triangle. If four vertices of the square are visible, then the square is above the triangle. Otherwise, the triangle is above the square. Thus, the 1st pile, the 4th pile, and the 5th pile are all piles with the triangle above the square.

14. (A) 1, 3, and 5

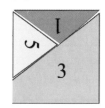
The bottom of 1 has the same length as the bottom of 3, so they can be opposite sides of the square. Fit 5 between the two pieces to form a square.

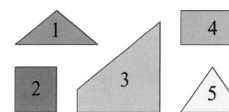

15. (C) 4
There is only one way to fill in the table. Here are the first steps:

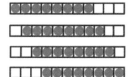

We now see that 1 and 3 need to be in the two shaded cells, so the sum is 4. The solution for the whole table is shown below.

1	3	2
3	2	1
2	1	3

16. (D) 5

Only the first three cells and the last three cells can be empty, so 11 − 3 − 3 = 5 cells must each contain a coin.

17. (D)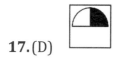

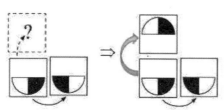

Each turn is like a mirror reflection. Turn over the card on the lower right on its left edge to see the picture of the original card in the lower left. Next, turn over the card in the lower left on its top edge to see the picture on the top.

18. (B) 27

If we subtract 3 from the unknown sum of the four brothers' ages, then we have a multiple of 4 since Paul's age − 3 is the age of the other three brothers. From the list of 25 − 3, 27 − 3, 29 − 3, 30 − 3 and 60 − 3 only 27 − 3 = 24 is a multiple of 4. Thus, 27 can be the sum of the ages of the four brothers: 6 + 6 + 6 + 9 = 27.

19. (D) 50

Each tree has twice as many pears as apples. Thus, any magic garden with 25 apples must have 2 × 25 = 50 pears. There are two types of magic gardens with 25 apples. We may have 7 trees with 3 apples each and 1 tree with 4 apples, or we may have 3 trees with 3 apples each and 4 trees with 4 apples each. In short, 7 × 3 + 1 × 4 = 25 and 3 × 3 + 4 × 4 = 25.
One garden is shown below.

20. (C) 6

Each dog has 3 more legs than noses, so 6 dogs have 6 × 3 = 18 more legs than noses.

21. (B) between bowl Q and bowl R

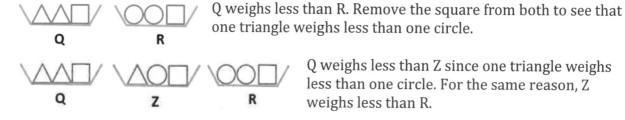

Q weighs less than R. Remove the square from both to see that one triangle weighs less than one circle.

Q weighs less than Z since one triangle weighs less than one circle. For the same reason, Z weighs less than R.

22. (C) 1917

The sum of the six numbers, without 201, is 2016 − 201 = 1815.
The sum of seven numbers, with 102, is 1815 + 102 = 1917.

23. (E)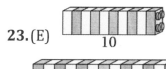

27 = 9 + 18, so the long bar is broken into bars of 9 and 18 bricks.

We cannot get the bar of 10 bricks from the bar of 9 bricks.

18 = 6 + 12, so the only chance to get a bar of 10 bricks is from the bar of 12 bricks, but 12 = 4 + 8, so we cannot get a bar of 10 bricks. We already know how to create bars of 8, 6, and 4 bricks. 6 = 2 + 4, so we also can get a bar of 2 bricks.

24. (B) Bertha

Originally, Angel chirps 4 times, Bertha once, Charlie twice, David 3 times, and Eglio 4 times.
If Angel turns, he would not chirp at all. If it was Bertha, she would chirp 3 times.
If Charlie turns, he would still chirp twice. If David turns, he would chirp only once.
If it was Eglio, he would not chirp at all.
So, only Bertha chirps more times when she looks in the opposite direction.
See the picture below.

Solutions for Year 2017

1. (E)

Notice that 8 – 3 is 5. That means that the puzzle we are looking for has to start with = 5. This eliminates answers (B) and (D). Also the second addition or subtraction must be equal to 2. This eliminates (A) and (C). The only puzzle piece fitting between the two given puzzles is puzzle (E).

2. (A) 12
   In the picture there are 6 kangaroos. John sees only half of the kangaroos. The other half must also be 6. The total number of kangaroos is 12 since 6 + 6 = 12.

   John sees half of the kangaroos. 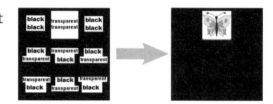 Half of the kangaroos are not seen by John.

3. (E)
   Any of the pictures cannot be seen if they are covered with at least one black square. Sliding both transparent grids with some dark squares would result in all but one square not covered with at least one black square. That square is in the top row, middle column, showing the butterfly. See the picture.

4. (C)
   All the sets of footprints but one from the first picture repeat in the upside-down picture. The picture highlighted in yellow does not appear in the upside-down image.

   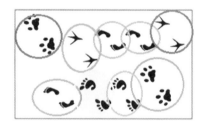 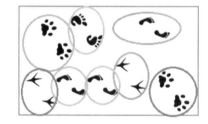

5. (A) 16
   Complete the addition and subtraction statements from left to right. 26 − 10 = 16 is the number hidden behind the panda.

   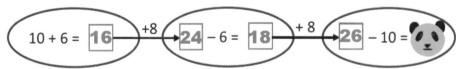

6. (E) 16

+	11	7	2
6	17	13	8
9		16	11

   The sums in the upper row are obtained by adding 6 to the numbers above the grid. They are 17 = 11 + 6, 13 = 7 + 6, and 8 = 2 + 6.
   The numbers in the lower row are obtained by adding 9 to the numbers above the grid since 11 = 2 + 9. The number in the box with the question mark is 7 + 9 = 16.

SOLUTIONS 2017

7. (C) 4
   There are 4 pieces that have exactly four sides. They are marked with red stars in the picture on the right.
   All remaining broken mirror pieces have either three or five sides.

8. (A)
   The position of the necklace in its original form is shown here:

   The position of the necklace after untangling the middle two beads is shown below.

   The position of the beads that are connected on the necklace is shown in (A):

9. (E)
   Notice that standing in the back of the house the chimney would be seen on the left side of the roof. Additionally, the house has three windows and no door. Houses (A) and (B) have doors. Houses (C) and (D) have chimneys on the right side of the house.
   Only house (E) has a chimney on the left side of the house, no door, and three windows.

10. (E) ● + ● = ■

    The statement ● + ● + ● + ● + ■ = ■ + ■ + ■
    can be simplified by removing one square from each side, so

    ● + ● + ● + ● = ■ + ■

    As indicated by parentheses below, there are two identical pairs on the left side and two identical squares on the right side,

    [● + ●] + [● + ●] = ■ + ■

    so one pair matches one square as shown here.

    Therefore, only statement (E) is true.

11. (B) 4

Getting three 25-balloon packets is more than 70 balloons since 3 × 25 = 75. Any other combination of 3 packets, such as two 25-balloon packets and one 10-balloon packet is not enough since 2 × 25 + 1 × 10 = 60 is less than 70 balloons. Marius can buy exactly 70 balloons by getting 4 packets, two 25-balloon packets and two 10-balloon packets, 2 × 25 + 2 × 10 = 70.

12. (C)

Folding the paper as shown in (C) produces the desired result:

Folding the paper in other ways will result in different patterns as shown below:

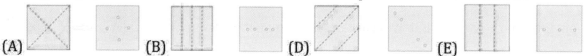

13. (D) 4

13 + 19 = 32 children signed up for the tournament was. 32 children cannot be divided into 6 teams evenly. 36 is the first number greater than 32 that is divisible by 6. At least 4 more children (36 − 32 = 4) need to sign up to form six teams with an equal number of members each.

14. (D) 14

There are three 2 × 2 squares if we look at rows 1-2 together. The same is true looking at rows 2-3 and at rows 3-4. The sums for the 2 × 2 squares are: 8, 5, and 7 in rows 1-2; 13, 12, and 8 in rows 2-3; 11, 14, and 9 in row 3-4. The largest sum is 14 that is obtained by adding 7, 3, 3, and 1.

15. (C) 75 min

The time needed to cook all 5 dishes is 40 + 15 + 35 + 10 + 45 = 145 minutes. Each cooking time is a multiple of 5, so the optimal times must be multiples of 5. The shortest cooking time would occur if the total cooking time of all 5 dishes could be equally divided among two burners. This cannot happen because 145 ÷ 2 = 72 1/2 and it is not a multiple of 5, so we must look for the total cooking time to be close to half the time needed to cook all 5 dishes and to be a multiple of 5. The optimal distribution of the total cooking times is 75 + 70 and it can actually happen. On one burner David can cook dishes with the cooking times of 40 and 35 minutes (75 minutes total) and on the second burner dishes with the cooking times of 15, 10, and 45 minutes (70 minutes total). Both burners would work simultaneously. The shortest time needed to cook all five dishes is 75 minutes.

SOLUTIONS 2017

16. (D) 13

Look at the diagram and the operations to be performed.
Any number multiplied by zero results in 0, so the number to the left of the question mark is 0.
Follow the arrows to perform all other operations.
The number 13 should be written in the circle with the question mark, which is marked in yellow here.

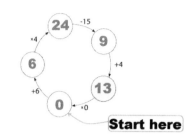

17. (C) 5

The plan represents the shape of the base of the building. The numbers represent how many blocks are put on top of each square space. There are 3 blocks shown by the upper square covered by ink. They are marked in red in the figure below. There are 2 blocks shown by the other square covered by ink, one shown in blue and one underneath. The sum is 3 + 2 = 5.

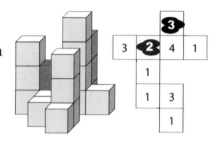

18. (B) 115 m

Overlaying the two given pictures gives a useful presentation of this problem.

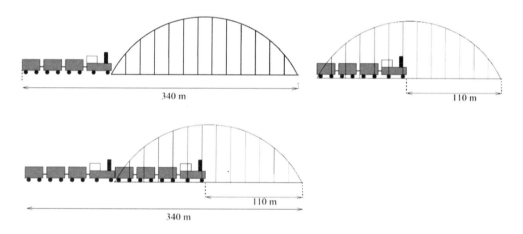

340 meters is the sum of the lengths of the bridge and the train.
When the train is on the bridge, the remaining length of the bridge is 110 meters.
As shown in the last picture, 340 meters is the sum of the lengths of two trains and 110 meters.
The length of two identical trains is 340 meters – 110 meters = 230 meters. The length of the train is 230 meters ÷ 2 = 115 meters.

19. (E) X

John was born in February and the problem asks about how old he is in March, so his birthday has already passed. He was born in the year MMVII. From the question's explanation, M=1000, the other M=1000, V=5, I=1 and I=1. Adding the numbers together we get:
1000 + 1000 + 5 + 1 + 1 = 2007. In order to calculate how old John was in 2017 it is enough to subtract the two given years. 2017 – 2007 = 10. John was 10 years old, which is X in Roman numerals.

© Math Kangaroo in USA, NFP   189   www.mathkangaroo.org

20. (D) 9

Susan can start her tour with the giraffe, the elephant, or the turtle since she does not want to start with the lion. She wants to see two different animals. After starting with the giraffe, she can then see the elephant, the turtle, or the lion. After starting with the elephant, she can then see the giraffe, the turtle, or the lion. After starting with the turtle, she can then see the giraffe, the elephant, or the lion. Thus, she can start with 3 different animals and then for each one she has 3 choices at to the second animal she sees. So, there are 9 different tours she can plan. The table below shows Susan's nine options to see two different animals.

21. (C) 5

Start with the fact that three brothers ate 9 cookies, so the fourth brother ate 11 – 9 = 2 cookies. Among the three brothers, one ate 3 cookies, so the two remaining brothers together ate 9 – 3 = 6 cookies. As a sum of two different numbers (ignoring the order) 6 is either 1 + 5 or 2 + 4. The fourth brother ate 2 cookies, so the only valid option left is 1 + 5 = 6. 5 was the largest number of cookies that one of the brothers ate.

22. (B) 5

In order to determine where the smileys are hiding, start with selecting the order of discovering cells with a smiley. Step 1 – RED arrows indicate the only options of 3 neighboring cells with a smiley; Step 2 – BLUE and PURPLE arrows are determined by Step 1, so there are no more smileys around two cells with 2; Step 3 – two GREEN arrows are already determined by Step 1 and there is only one option for the third smiley; Step 4 – for the BROWN arrow there is only one option left in the lower left corner. The cells left shaded cannot have smileys. Zosia hid 5 smileys as the diagram below shows.

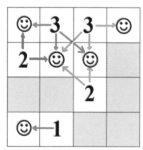

23. (E) 19

The total number of pieces of candy in the bags is 1 + 2 + 3 + 4 + 5 + 6 + 7 + 8 + 9 + 10 = 55. The first four boys took 5 + 7 + 9 + 15 = 36 pieces of candy, so Eric got 55 − 36 = 19 pieces of candy.

24. (B) 2

The total number of petals is 6 + 7 + 8 + 11 = 32 = 30 + 2. After 10 rounds of tearing off one petal from each of any three flowers (if it can be done), Kate will stop with 2 petals left. It can be done in many different ways. One is shown in the picture below.

Solutions for Year 2018

1. (B) 6

Anytime a digit is repeated, the same stamp is used. So, only count the number of different digits in the date.

2. (E) 6

The picture shows the path of the arrows. The balloons which each arrow hits are marked with an X.

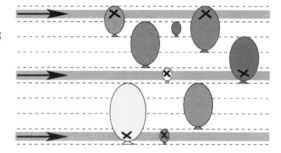

3. (C) 18

Susan	sister	brother
6 years old	(6 − 1) years old	(6 + 1) years old

6 + 6 + 1 + 6 − 1 = 6 + 6 + 6 = 18

SOLUTIONS 2019

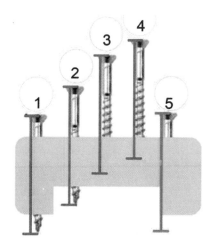

4. (E) 5
   If screw 5 was as long as any of the other screws, the bottom would be visible under the block like it is for screw 1. Therefore, screw 5 is the short screw.

5. (D)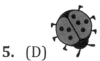
   Sophie has 3 dots on her left wing and 4 dots on her right wing. The ladybug in answer choice (D) has 4 dots on her left wing and 3 dots on her right wing. Therefore, (D) shows a different ladybug. The other four answer choices match the dots Sophie has.

6. (D)
   The folded sheet of paper has a large opening that looks like a half-circle. If the paper is opened and this is the middle, there will be a hole resembling a circle in the middle of the sheet.

7. (D) 21
   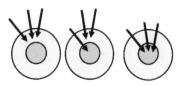
   12 points = 4 + 4 + 4
   15 points = 4 + 4 + 7
   7 + 7 + 7 = 21 points
   On the first turn, each of the arrows hit an area with the same color, so Diana received the same number of points for each arrow. Since there were three arrows, Diana scored 4 points with each of them (4 + 4 + 4 = 12). On the second turn, one of the arrows landed on an area worth more points. Diana scored 4 + 4 = 8 points with the two arrows on the lighter blue area, so the darker blue area is worth 15 − 8 = 7 points for each arrow that lands there. So, on the third turn, Diana scored 7 + 7 + 7 = 21 points.

8. (A) 5
   Going around the table, think of the right and left side as a person facing it from that direction would see it.
   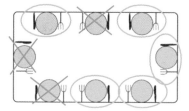

9. (D) 4

The original tile has one circle and one star, so any design must have an equal number of circles and stars. The design shown second from the left has six circles and eight stars, so Roberto definitely cannot make it. Below are suggestions how the other designs can be made.

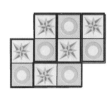 

10. (A)

The two figures missing in the fourth column are the rhino and the ghost. If Albert places the rhino in the spot with the question mark, then he would have to place the ghost in the only other empty spot in the fourth column, making two ghosts in the second row, which is not allowed. Therefore, the ghost must go in the spot with the question mark.

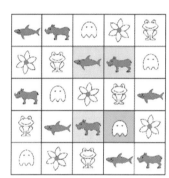

11. (B) 6

Tom needs to use the squares for the top part of the boat because of the right angles. Three squares will fill the top part of the boat. Three trapezoids will make up the bottom of the boat if two are flipped.

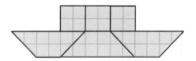

12. (E)

Only choices (C) and (E) have the right number of dots. However, in (C) the black dots are too close to the one that is outlined.

13. (E) Friday

Twelve carrots take 6 days to eat. Work backwards with the days of the week.
11th and 12th carrots: Wednesday
9th and 10th carrots: Tuesday
7th and 8th carrots: Monday
5th and 6th carrots: Sunday
3rd and 4th carrots: Saturday
1st and 2nd carrots: Friday

14. (C) 8

Each cube has 6 faces. Where it touches another cube, a face is covered, and will not be painted. So, any cube that touches exactly two other cubes will have exactly four faces exposed and painted. This is true for all cubes in this figure except for two, one at each end of the structure. So, 10 − 2 = 8 cubes will be painted on exactly 4 faces.

15. (C) 4

    The number of butterflies on the flowers is twice the number of dragonflies on the flowers, so the total number of insects on the flowers is 3 times the number of dragonflies on the flowers. That total is between 5 and 8 since the total is more than half or 8, which is 2, and at most 8 since no more than 1 insect is on each flower. The only multiple of 3 between 5 and 8 is 6, so there are 6 insects and one-third of them (which is 2) are dragonflies. Hence, 6 – 2 = 4 insects are butterflies.

16. (D) 33 km

    Because the distance between Easter and Flower going through Volcano is 26 km, we can disregard the information that the distance between Easter and Volcano is 17 km. Now we can subtract the two distances of 26 km each and the 15 km between Flower and Desert from the 100 km round trip to find the distance between Easter and Lake.
    100 – (26 + 15 + 26) = 100 – 67 = 33.

17. (D) D

    Two paths ending at the door D are shown. There is no path ending at any other door. Otherwise such a path could be reversed and the reversed path would show decreasing numbers. Starting at A we could move from 12 to 9 to 5 (in two ways) and not farther. Starting at B we could move from 14 to 12 to 9 to 5 (in two ways), from 14 to 5, or from 14 to 6 and not farther. Starting at C there is no way to move anywhere from 6. Finally, starting at E there is no way to move from 9.

    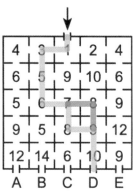

18. (C) C

    According to the second scale, C weighs **at least** 10 g + 20 g = 30 g, so C and D together weigh at least 30 g + 10 g = 40 g. According to the first scale, C and D together weigh less than half of all four balls together. The four balls weigh 100 g since 10 + 20 + 30 + 40 = 100, so C and D together weigh less than 50 g. Hence, C and D together weigh **at most** 40 g.
    Therefore, C and D together weigh **exactly** 40 g, C alone weighs 30 g, and D alone weighs 10 g.
    Notice that B weighs 30 g – 10 g = 20 g and A weighs 40 g.

19. (B) 8 cm

    Tightening the band by one hole makes it 2 cm shorter. It does not matter that there are two layers in the overlap, because you can think of the lower layer as not moving. Moving from one hole to five holes decreases the length by 4 × 2 cm = 8 cm.

20. (A)

The result of adding the same number four times can be 4, 8, 12, 16, ...
Among the numbers 1, 2, 3, 4, 5 only 4 is a possible result, so

☙ = 4, ☼ = 1, and ⚛ = 2. 1 + 4 = 5, so = 5.

Hence, represents the number 3.

21. (B)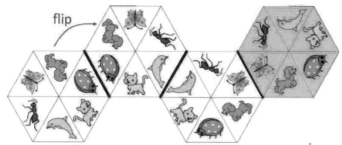

22. (D) 198
The smallest squares have an area of 1, and therefore a side length of 1, because $1 \times 1 = 1$. Knowing this, we can find the side lengths of all the squares that make up the large rectangle. The side lengths of the rectangle are 18 and 11, so the area is 198. We can also add the areas of all the squares to find the area of the rectangle.

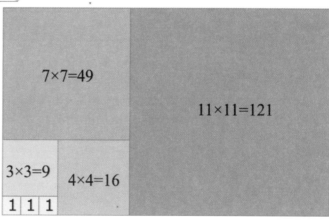

23. (E) only the numbers 1 and 7
The cell with the question mark has 5 neighbors, and there is only one cell that is not a neighbor with it. A number in the sequence that cannot be placed next to each other will be one more or one less than a given number. Because there are 7 numbers, the only numbers that can be written in the cell with the question mark are the numbers either at the start or at the end of the sequence, because they have only one other number in the sequence that cannot be placed next to them. These are the numbers 1 and 7. Placing any other number in the cell with the question mark will violate the rules because then consecutive numbers will necessarily be neighbors with it.

24. (B) 9
Each time 3 heads are cut off, 1 new head grows. 13 heads = 4 × 3 heads + 1 head, so the dragon grew 4 **new** heads. Together with the original heads, there were 13 heads cut off. Therefore, 13 − 4 = 9 is the number of heads the dragon had at the beginning.

Solutions for Year 2019

1. **(E) E**
   The runner with D on the shirt stands on the highest step. The next one is B and the third one is E.

2. **(C)**
   12 is 10 + 2, which is the same as 5 + 5 + 1 + 1, so the picture will have two bars and two dots. This is picture (C).

3. **(A) Tuesday**
   Since yesterday was Sunday, then today is Monday and tomorrow will be Tuesday.

4. **(D)**
   Olaf sees the second, fourth, and fifth vehicle (counting from the left) when he closes his book. The distances from the book spiral to the first opening is 3 squares, which will cover up the red car. The two by two opening would show the blue motorcycle. The distances from the book spiral to the second opening is 7 squares, so the next vehicle, which is the green truck, is covered. The four by two opening would show the orange SUV and the purple tractor. This is shown in answer (D).

5. **(A)**
   The only piece that Karina can get is the one with a star and a club shown in the middle of the top row. All other pieces are either not next to each other or the shapes don't face the right direction.

6. **(A)**
   The track of shoeprints in the straight line from right to left is covered by the other two. This means this was the first person walking on the snow. The second track, which covers the first track but is covered by the third, starts in the lower right corner goes up and then down. The third is the track that is on top of the other two, moving from the upper right corner to the middle left. This order is shown in answer (A).

7. **(D)**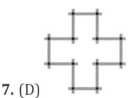
   There are 10 connected sticks in Pia's shape shown in the picture. Answers (A), (B), (C), and (E) also show 10 connected sticks. Answer (D) shows a shape with 12 connected sticks, which is more than Pia has.

8. (B) 5

Starting with 2 + 1 = 3, 0 + 3 = 3, and 1 + 8 = 9 (answers shown in blue), we can calculate 8 − 3 = 5 (shown in red), which is the replacement for the question mark.

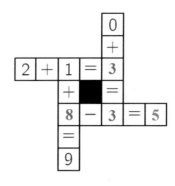

9. (B) 16

Linda uses 8 pins to pin 3 photos on a cork board as shown in the picture. She uses four pins for the first photo, but because of overlapping edges, she only needs two more pins for each additional photo. When pinning 7 photos, Peter would use 16 pins (4 + 6 × 2 = 4 + 12 = 16), as shown below.

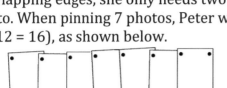

10. (C) 3

By removing one square, marked in gray, only the first, second, and third shape can be obtained. So, Dennis can get 3 out of 5 listed shapes.

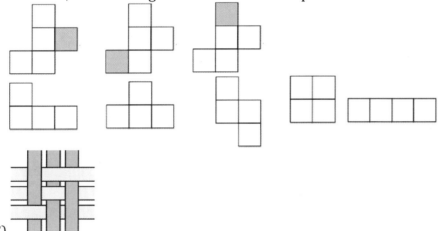

11. (C)

From the back, the middle vertical strip will be behind the upper two horizontal strips and the third horizontal strip will be behind the middle vertical strip. Only picture (C) shows this. Also, the left vertical strip of (C) matches the right strip of the original pattern when looking from behind. The same is true for the right vertical strip of (C) and the left vertical strip of the original pattern.

12. (E) 11 kg

From the first scale we know the toy dog's weight is less than 12 kg. From the second scale we see that 2 dogs weigh more than 20 kg, which means one dog weighs more than 10 kg. There is only one whole number that is greater than 10 but less than 12, and it is 11.

13. (B) 10

Sara will first need to trade blue marbles for red marbles, and then red marbles for green ones.

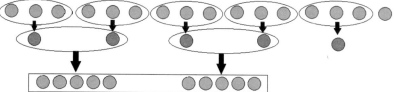

14. (A) Either 0 or 1

In order to get the largest possible sum, Steven needs to put the greatest digits first, so the hundreds digit is 9 and the tens digit is 2. Steven will add the last digit of the 3-digit number to the other remaining digit, so it does not matter whether he places 1 or 0 there. The sum will be the same.

15. (D) 250

Since the glass full of water weighs 400 grams and the empty glass weighs 100 grams, the water alone weighs 400 – 100 = 300 grams. A glass half-full would weigh 100 grams (glass only) + 150 grams (half of 300 grams of water), which gives the sum of 250 grams.

16. (D) 11 cents

By combining all three costs and rearranging the fruit we can conclude the following:

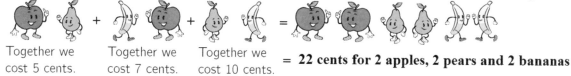

If 2 apples, 2 pears, and 2 bananas cost 22 cents, then 1 apple, 1 pear, and 1 banana cost half of this, which is 11 cents.

17. (E) 6

Starting with the middle row we can calculate that the circle stands for 4, as 12 divided by 3 is 4. This gives us the following:

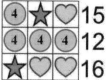

We can subtract 4 from the first row to get the following:

Notice that one more heart in the last row makes the sum larger by 5, so the heart stands for 5. Now we know that the star stands for 6, because 6 + 5 = 11 and 6 + 5 + 5 = 16.

SOLUTIONS 2019

18. (C) 44

Anna used a set of 7 squares along each of the four sides of the picture and added four more squares at each corner. (They are colored in the picture to the right.)
A picture that is 10 by 10 requires the frame to have 10 squares along each side plus one in each of the four corners.
10 + 10 + 10 + 10 + 4 = 44 squares

19. (B) 64

From 1 to 49 the digit 5 appears 5 times and from 50 to 59 it appears $10 + 1 = 11$ more times. Thus, from 1 to 59 the digit 5 appears $5 + 11 = 16$ times as given. We can have more pages included before the digit 5 occurs for the 17th time on page 65.
Thus, the maximum number of pages this book can have is 64. The digit 5 that appears 16 times is marked in red and the last number represents the maximum number of pages.

1, 2, 3, 4, **5**, 6, 7, 8, 9, 10, 11, 12, 13, 14, 1**5**, 16, 17, 18, 19, 20, 21, 22, 23, 24, 2**5**, 26, 27, 28, 29, 30, 31, 32, 33, 34, 3**5**, 36, 37, 38, 39, 40, 41, 42, 43, 44, 4**5**, 46, 47, 48, 49, **5**0, **5**1, **5**2, **5**3, **5**4, **55**, **5**6, **5**7, **5**8, **5**9, 60, 61, 62, 63, **64**.

20. (E) 83

The horizontal length of the path is equal to the sum of 36 and 28, which is 64. The red line in the picture marks the horizontal path. The vertical length of the path is equal to the sum of (20 – 4) and (6 – 3), which is 16 + 3 = 19. The blue line in the picture marks the vertical path. The cat walks 64 + 19 = 83 meters.

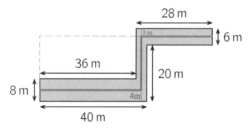

21. (B) 3

Since 10 of the animals are not cows, we can conclude that there are 5 cows, since there are 15 animals total. Since 8 are not cats, we can conclude that 15 – 8 = 7 are cats. This leaves us with the number of kangaroos in the park as 15 animals – 5 cows – 7 cats = 3 kangaroos.

22. (E) 1 and 3 are yellow

Triangle 4 must be blue as it has an edge common with red and yellow. Triangle 5 must be red as it has an edge common with yellow and blue.

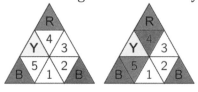

For triangles 1, 2 and 3, Mary must use the remaining triangles, two of which are yellow and one is red. Since two triangles of the same color cannot share an edge, the yellow ones must be 1 and 3. This leaves us with only one option, red, for triangle 2.

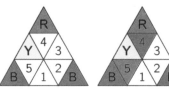

Answer "(E) 1 and 3 are yellow" is the only one that correctly describes the figure.

23. (B) Bartek

If Edek is telling the truth, then Alek ate the cookie, and Alek would be a liar by contradicting Edek. There is only one liar among the five boys, so Bartek's statement would be true. Bartek says that he (not Alek) ate the cookie. Therefore, Edek is a liar and everybody else is telling the truth, so Bartek ate the cookie.

24. (D) 22

At the beginning hang all 35 towels as in figure 2. The number of pegs still available is $58 - (1 + 35) = 22$ since each towel has one peg at its right end (35 pegs) and the leftmost towel also has one peg at its left end. One by one, we move 22 consecutive towels to the left. Each of these 22 towels already has a peg at its left end and we add a new peg to the right end of each towel. Therefore, Emil hung up 22 towels as shown in figure 1.
Notice that $35 - 22 = 13$ towels were hung up as shown in figure 2.
Altogether, $22 \times 2 + (1 + 13) = 44 + 14 = 58$ pegs were used by Emil.

Part III: Answers

Answer Keys

	1998	1999	2000	2001	2002	2003	2004	2005
1	C	C	B	C	B	E	E	C
2	D	E	C	D	C	C	A	A
3	C	B	D	D	D	D	C	B
4	D	C	B	C	D	A	C	B
5	C	D	B	B	D	C	E	B
6	C	E	C	E	C	D	B	D
7	D	C	B	E	A	C	E	E
8	C	B	C	E	B	C	B	C
9	E	E	B	C	C	C	D	C
10	C	B	D	D	E	C	D	B
11	C	C	B	E	C	D	B	B
12	A	A	D	E	A	A	B	C
13	B	D	B	C	E	C	E	D
14	D	A	C	B	B	A	B	D
15	B	D	C	C	C	B	E	E
16	D	D	C	A	E	C	C	D
17	E	D	B	C	A	D	E	B
18	D	C	B	C	B	E	C	B
19	B	C	E	D	D	E	C	A
20	B	A	C	A	E	A	A	D
21	B	A	C	A	C	B	B	C
22		D	A	B	D	D	B	E
23		C	E	D	E	A	E	B
24		A	A	E	D	B	E	E

ANSWER KEYS

	2006	2007	2008	2009	2010	2011	2012	2013
1	C	C	C	E	D	A	B	D
2	B	A	D	C	C	C	D	D
3	D	C	B	B	C	C	B	E
4	A	C	B	A	C	E	C	B
5	D	C	D	B	D	B	A	C
6	E	E	A	B	A	A	E	D
7	B	C	E	C	C	B	B	B
8	B	B	A	D	B	D	D	E
9	C	C	E	C	E	C	E	B
10	A	C	D	D	B	B	B	D
11	A	A	E	A	B	B	E	D
12	D	C	D	B	D	C	C	A
13	C	A	E	B	D	E	D	D
14	E	B	C	D	C	A	B	E
15	A	B	B	B	D	D	D	B
16	C	A	B	A	A	C	C	B
17	B	E	C	E	E	A	D	D
18	D	B	A	D	C	C	C	D
19	D	C	D	B	D	C	D	B
20	E	A	E	D	D	C	B	C
21	B	A	E	A	E	E	E	E
22	E	B	B	B	B	D	D	B
23	C	D	C	E	D	A	E	B
24	B	E	D	B	E	E	C	B

ANSWER KEYS

	2014	2015	2016	2017	2018	2019
1	D	E	E	E	B	E
2	D	A	E	A	E	C
3	A	E	A	E	C	A
4	D	C	A	C	E	D
5	A	E	D	A	D	A
6	E	E	B	E	D	A
7	E	B	A	C	D	D
8	C	D	C	A	A	B
9	E	A	B	E	D	B
10	E	D	B	E	A	C
11	B	B	C	B	B	C
12	D	B	B	C	E	E
13	B	C	D	D	E	B
14	B	B	A	D	C	A
15	C	D	C	C	C	D
16	B	C	D	D	D	D
17	B	A	D	C	D	E
18	C	C	B	B	C	C
19	B	A	D	E	B	B
20	C	D	C	D	A	E
21	A	E	B	C	B	B
22	D	C	C	B	D	E
23	D	D	E	E	E	B
24	D	C	B	B	B	D